D0204049

Environmental Life Cycle Analysis

Environmental Life Cycle Analysis

David F. Ciambrone
Hughes Aircraft Company
Newport Beach, California

Lewis Publishers

Boca Raton   New York

Acquiring Editor:	Joel Stein
Project Editor:	Andrea Demby
Marketing Manager:	Arline Massey
Typesetter:	Pamela Morrell
Cover design:	Denise Craig
PrePress:	Carlos Esser
Manufacturing:	Carol Royal

Library of Congress Cataloging-in-Publication Data

Ciambrone, David F.
    Environmental life cycle analysis / David F. Ciambrone.
        p. cm.
    Includes bibliographical references and index.
    ISBN 1-56670-214-3
    1. Industrial ecology.   2. Product life cycle—Environmental
aspects. I. Title.
TS161.C497   1997
658.5′752—dc21
                                         97-8058
                                            CIP

No claim to original U.S. Government works
International Standard Book Number 1-56670-214-3
Library of Congress Card Number 97-8058
Printed in the United States of America 1 2 3 4 5 6 7 8 9 0
Printed on acid-free paper

PREFACE

This book is designed to assist design/systems engineers and managers in determining how to best design or change a product or set of processes to minimize the impact on the environment over the life cycle of the product or process. The first two chapters provide an introduction and overview of the environmental life cycle analysis process.

Chapter 3 describes the basic framework, definition of boundaries, the use of checklists, data gathering issues, construction of models, and interpretation of results. This chapter establishes the basis and methodologies required.

Chapter 4 defines templates and special cases that may be encountered and concepts for handling them. Chapters 5 through 9 go into detail about the modeling, issues, and data collection for each stage of the product life cycle. Chapter 10 provides a summary of the various steps and ideas on how to present data and reports.

The organization of the book follows a logical path for the user to understand the overall concept of environmental life cycle analysis and to walk through each of the stages. The stages are broken down into individual topics to provide a clearer picture of what is included in each stage and the issues involved. The book tries to provide alternatives to the analysis process, thereby allowing the user greater flexibility while still gaining constructive information.

ABOUT THE AUTHOR

Dr. David Ciambrone is an energetic manager, management and environmental consultant, professor, lecturer, writer, and inventor, with more than 30 years experience in the manufacturing, business, and environmental fields. His main fields of interest include environmental management, manufacturing, waste minimization, and wastewater treatment.

Dr. Ciambrone has been Corporate Director of Environmental Affairs for a Fortune 100 company, Director of Manufacturing Engineering and Acting Vice President of Manufacturing for Fortune 500 companies, as well as Manager of Engineering and Business Development for a major environmental company.

He has designed numerous manufacturing processes used in the electronics and metal finishing fields. Dr. Ciambrone has established and managed surface mount assembly operations, hybrid microelectronics facilities, circuit board fabrication plants, and industrial plating operations. He has extensive experience as a turnaround manager for major companies and has experience in the implementation of ISO 9001. Dr. Ciambrone has also managed facilities engineering organizations for a Fortune 100 company. He has designed water and wastewater systems for use on land and by the marine industry for companies in the U.S. and abroad. Some of the systems used novel approaches to waste treatment. He has been a principal consultant in environmental companies conducting engineering projects and business development in the U.S. and internationally.

Dr. Ciambrone served on the Board of Directors of an international environmental company and served on the Management Oversight Board of the Resolution Trust Corporation (U.S. Treasury). He was also President of the southern California chapter of the National Association of Environmental Managers and is a Fellow of the International Oceanographic Foundation. He is a long-standing member of the American Electroplaters and Surface Finishers Society (past President) and is a past President of a chapter of the

National Management Association. He is a Registered Environmental Assessor in California.

Dr. Ciambrone has numerous publications in the fields of metal finishing, waste treatment, and waste minimization and has published books on waste minimization and life cycle analysis. He has patents in the fields of wastewater treatment, chemical and biological weapon destruction, and industrial processes.

Dr. Ciambrone has taught courses in industrial and manufacturing engineering, materials science, business and operation management, TQM, and Just-in-Time methods at major universities. He has lectured at many technical and business conferences and universities on manufacturing. In addition, he has lectured on environmental subjects nationally and internationally.

Dr. Ciambrone received his Bachelors Degree from San Diego State University and his Doctor of Science from Brantridge College, University of Sussex in England. Sir David is also the Laird of Parton in Scotland.

ACKNOWLEDGMENTS

The author wishes to acknowledge the following individuals for their contributions in the preparation of materials for this book:

- Ms. Joan Moscrop, without her kind help, this book would not have happened
- Mr. James Waylonis for preparation of figures
- Mr. Norm Sealander, Northrop Grumman Corporation, for his support and encouragement
- Dr. D. Mehta, Hughes Aircraft, for his support and encouragement
- Kathy Ciambrone for her support and encouragement
- Ms. Andrea Demby, for her assistance and editorial support

CONTENTS

1   INTRODUCTION

The trend in U.S. industry, and the U.S. EPA, is toward preventing wastes before they are created instead of treating or disposing of them later. An example of the interest by the government is the U.S. Army Strategic Defense Command regulation USASDCR 200-2. This regulation requires an environmental impact analysis of weapon systems from design through use and disposal. The advancement of ISO 14000 is another example of companies, technical associations, and governments trying to organize a voluntary approach to waste prevention, In international markets this is extremely important. European countries have "green" requirements on products and are expanding the requirements at a rapid rate. This is evident in Germany, The Netherlands, Denmark, Scandinavian countries, and Great Britain. Asian countries, such as Singapore, have also initiated a similar set of programs. Part of a green product and green manufacturing is understanding the *total* impact and cost the product has on the environment. This means evaluating the total life cycle of a product from raw materials and manufacturing through distribution, use, and ultimate reuse/disposal (Figure 1-1). That a product has a finite life cycle implies that there are associated costs that go along with it. Examples of these costs are shown below:

- Design costs
- Stocking/handling costs
- User/operating costs
- Disposal costs
- Manufacturing costs
- Shipping/transportation costs
- Reuse/recycle costs
- Compliance/licensing costs

This illustrates that the impact on the environment and life cycle costs are larger than just the manufacturing costs (Figure 1-2). Understanding the life cycle costs will assist the company in finding ways to reduce postmanu-

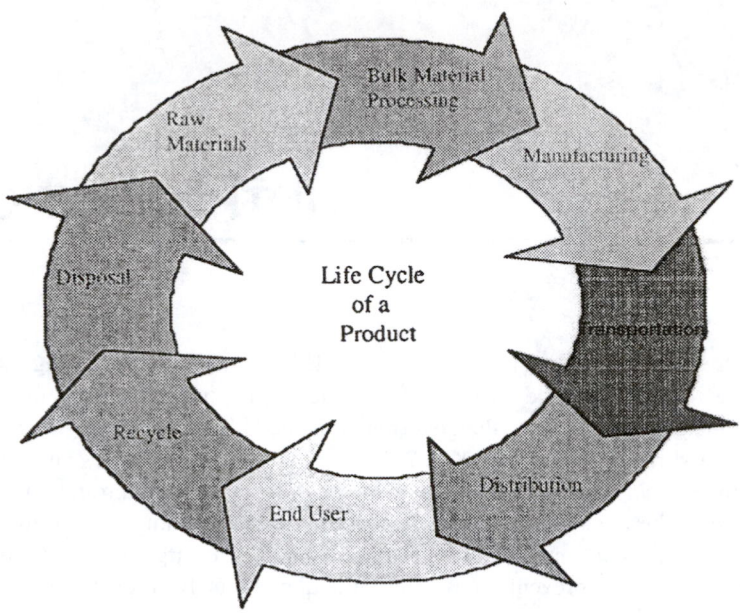

Figure 1-1.   Environmental life cycle analysis stages.

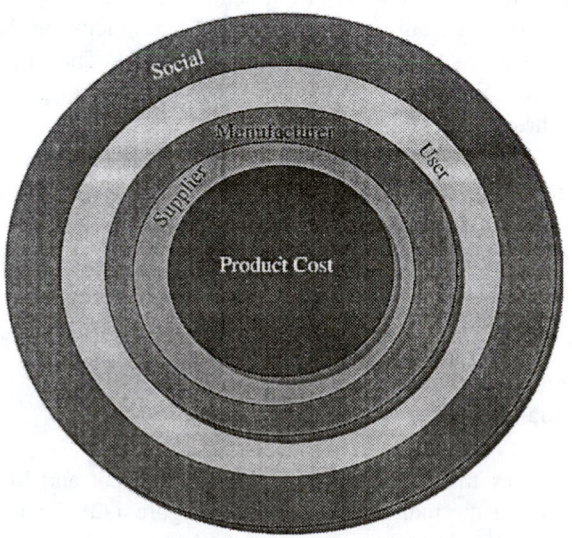

Figure 1-2.   Life cycle cost circle.

Figure 1-3. **Team approach to environmental quality and customer satisfaction. (From *Waste Minimization as a Strategic Weapon*, Ciambrone, D. F., CRC/Lewis Publishers, Boca Raton, FL, 1995.)**

facturing costs. Reducing these costs will also help in reducing the environmental impacts. While the concept of life cycle has existed for more than 20 years, there has been no comprehensive procedure that will facilitate understanding the process and provide the basic tools needed for a company to conduct the analysis in a rigorous yet cost-effective manner. This book would provide the industrial user with the necessary background, simple and effective tools, instruction, and examples to conduct environmental life cycle analysis on their products in a cost-effective manner.

Environmental life cycle analysis is a systematic tool used for assessing the environmental impacts associated with a specific product or service. The application of the process, and associated waste minimization practices by management, design, and manufacturing (Figure 1-3) can also lead to better and less-polluting products that are less expensive and provide a marketing edge over the competition. The importance of the life cycle analysis goes beyond the marketability of the product being studied. In Chapter 10, it will be shown that the outcome of the life cycle analysis can have a positive impact on human health, the ecosystem, and natural resources (Figure 1-4).

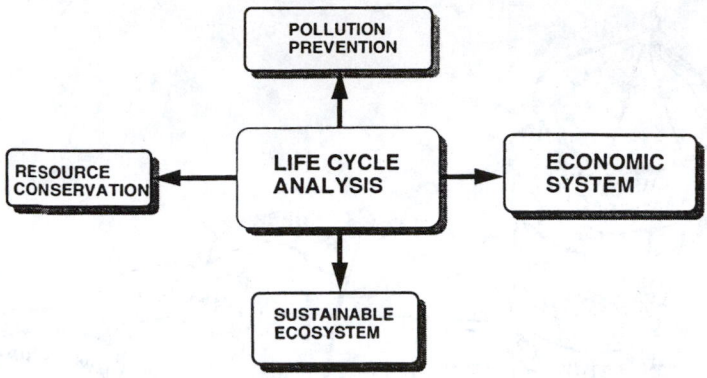

Figure 1-4.   Life cycle goals.

2   OVERVIEW

BRIEF HISTORY

Environmental issues have gained greater public and legal scrutiny in the past 25 years. The public has become more aware and interested in the consumption of goods and services and their impact on the natural resources and the quality of the environment. A nationwide study in 1991 by the *Wall Street Journal*/NBC found that 80% of Americans considered themselves environmentalists. In the past 12 years, an increasing number of firms have been shifting their focus from remediation of pollutants to pollution prevention. This has included "green" design activities and replacing some materials and manufacturing processes with more environmentally friendly ones. These efforts on behalf of the companies involved have drastically reduced the levels of industrial pollution. The impact, or "MANPRINT", on the environment by the products use and disposal has not usually been included in the environmental assessment of the pollution prevention program.

Industry is realizing that the impact their products have on the environment does not start and end with the manufacture of the product. The impact a product has on the world starts with the design and ends at the ultimate disposal of the product after its useful life. What considerations are made during the design phase of a product affects the whole life cycle. It is important to not have solely a means of determining the environmental impacts of the manufacturing process but what impact the product will have on the environment and its ability to be recycled.

While reducing the impact a company and its products make on the environment is a good idea, the costs associated with minimizing these impacts must be considered. Everything a company does has an effect on its bottom line. Environmental compliance has been considered a cost of doing business, and staying out of jail. The increased use of waste minimization programs has not only reduced the costs of compliance (which goes up over time) but has lowered the costs of producing the products. In a number of cases the quality of the product has improved. This in itself reduces costs and improves competitiveness of the products. In the global marketplace, the reduced impact of products on the environment has serious marketing impact. Sound environmental practices will result in designs that meet or exceed the requirements of the countries where they will be sold. This is especially true in Europe where manufacturers retain responsibility for disposal of their products after retired by their users. There are laws on the books in Germany that require the removal of unnecessary packaging from products and lets merchants pay for the disposal of this waste. It is reasonable therefore to consider the total impact on the environment of a product and plan means of reducing that impact during the planning and design phase of the product life cycle.

Life cycle analysis had its beginnings in the 1960s. One of the first publications on the subject was presented at the World Energy Conference in 1963 by Harold Smith. He reported on his calculation of cumulative energy requirements for the production of chemical intermediates and product. Later in the 1960s global modeling studies were published in *The Limits to Growth* (Meadows et al., 1972) and *A Blueprint for Survival* (Club of Rome). These studies involved the demand for finite raw materials and energy resources. In 1969 researchers did a study for The Coca-Cola Company that was to be the foundation for current methods of life cycle analysis. In the early 1970s other companies in the U.S. and Europe performed similar life cycle analyses. The process of quantifying the use of resources and the release of products into the environment in the U.S. became known as Resource and Environmental Profile Analysis (REPA). In Europe it was called Econobalance. During the late 1970s to early 1980s interest in REPA waned in the U.S. However, interest in the process increased in Europe. The Green Movement, especially the formation of Green political parties in Europe, reawakened interest in the subject. When solid waste became an issue worldwide in the late 1980s, the life cycle analysis technique reemerged as a tool for analyzing environmental problems. The Iraqi invasion of Kuwait in 1990 and once again the prospect of oil shortages as well as concerns for solid waste disposal reemphasize the need for environmental life cycle analysis tools.

The concept of a life cycle as used here simply means that the inputs to the "cycle" (energy, materials, etc.) and outputs (energy, waste materials, products, etc.) are evaluated for each step of a product or process life. The cycle begins at conception of the product or process and completes with the recycle/disposal of the product and its constituents (Figure 2-1). In the past, companies have used the "cradle to grave" analysis as indicated. During the 1995 to 1997 time frame, some companies have used the "fence line approach". This approach assesses the impacts associated with the product on site and disregards off-site functions. While this approach does not provide valuable information needed to assess the total impact, it will provide insights into the operations at the site and their impact on the environment. It can also point out areas for environmental and cost improvements. The fence line approach is not a true life cycle analysis but should be considered when determining the exact requirement of the analysis. During a life cycle analysis, the impact on the environment is considered. From this one may be able to ascertain if

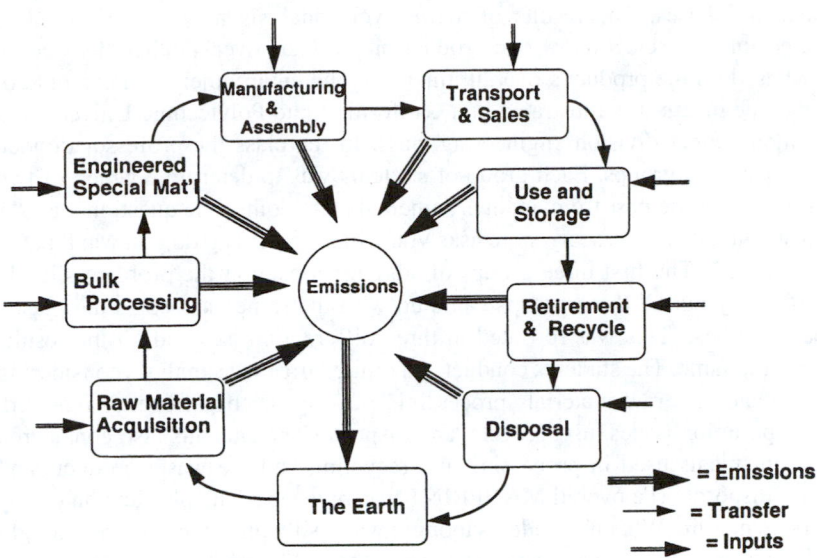

Figure 2-1.   Life cycle.

materials or processes may be used to minimize the impact the product has on the environment. This we refer to as the product's MANPRINT.

ASSESSMENT METHODOLOGY OVERVIEW

The environmental life cycle analysis methodology has three basic components. These components overlap and build on each other to develop a life cycle analysis. The three components are

- Inventory analysis
- Impact analysis
- Improvement analysis

It should be noted that the life cycle analysis can be conducted for use both internally to an organization as well as externally. A life cycle analysis that is not properly set up will cost far more than is necessary and be totally impractical. Setting the boundaries for a life cycle analysis is one of the most critical steps in conducting the analysis. Setting the proper boundaries will define what one will obtain from the study and determine the costs associated with the analysis.

A life cycle assessment inventory will provide a quantitative catalog of inputs (materials, energy) and outputs (including environmental releases) for a specific product, activity, or process. Once the inventory analysis has been completed and verified, the results can be used for the impact and improvement analysis phases. The results of a life cycle analysis may also be used to determine the selection of one product or products over another. This could be based on the product's overall impact on the environment. An example of this type of use was illustrated in a California State Polytechnic University at Pomona upper division engineering class. In the class the professor divided the class into groups. Each group of students was to determine which of two choices was the best from an environmental standpoint. The question was, "A checkout lady in a grocery store asks you what type of bag do you want, paper or plastic?" The first three groups of students looked at the problem slightly differently but all three groups came up with the same answer — the plastic bag was best. This was repeated in three different classes and all the results were the same. The students conducted a limited life cycle analysis considering raw materials, raw materials processing, making the bags, energy required, transportation issues, use, recycle, and disposal. The students also considered the chemicals used in processing and recycling and the waste products and their disposal. The overall MANPRINT was reviewed. The plastic won out by a large margin. What the students thought was a silly question (they all thought the answer would be paper because it's natural) but learned through an analysis was that the true solution was plastic. Life cycle analysis can change accepted thought and "truth".

The cataloging of inputs and outputs must have an objective. The specific objectives for conducting an inventory include, but are not limited to, the following:

- Compare inputs and outputs associated with alternative products, materials, or processes
- Determine points within the life cycle or given process where the greatest reduction in resource requirements and fugitive emissions might be achieved
- Assist in guiding the development of new products that reduce the overall MANPRINT
- Establish a comprehensive baseline
- Assist in training personnel responsible for reducing the environmental impact associated with products or processes
- Assist in the substantiation of product claims about reducing their impact on the environment
- Supply information for the evaluation of public policy affecting resource uses, recycling, or releases

LIFE CYCLE INVENTORY CRITERIA

It is recognized that inventories will vary from screening to extremely technical and practical as possible. To be a useful tool, however, a life cycle inventory should meet, as a minimum, the following criteria:

- Quantitative — All data should be quantified and documented with suitable quality control. Any assumptions in the data and methodology must be specified.
- Replicable — The sources of the information and methodology are sufficiently described or referenced so the same results could be obtained by a skilled person and evidence would be available to explain any deviations.
- Scientific — Scientific based analysis is used to obtain and process the data.
- Comprehensive — All significant energy and materials uses and waste releases are included. Any elements missing because of data unavailability or cost and time constraints are clearly documented.
- Detail — The inventory is conducted in a manner and to a level of detail commensurate with the purpose of the study.
- Peer reviewed — If the study results are to be used in a public manner, they should be peer reviewed using accepted protocols.
- Useful — The users of the study can make appropriate decisions in areas covered by the inventory. Any limitations regarding the utility of the study should be clearly noted.

The meeting of these criteria should make the life cycle inventory report useful, technically supportable, unbiased, and applicable.

EXAMPLES OF APPLICATIONS

In Chapter 1, the study of paper vs. plastic shopping bags was mentioned. In the analysis, conducted by students at the California State Polytechnic University at Pomona, the plastic bag won out over the paper bag as the most environmentally friendly. The studies considered:

- Raw materials processing
- Energy usage
- Bag manufacturing
- Usage
- Disposal and recycling

The students considered the chemicals used in the various processing steps and their MANPRINT. Table 2-1 shows the environmental impact resulting from the manufacture of one air dry ton of wood pulp by wood type. Figure 2-2 illustrates the impact on a landfill of disposal of 1000 paper and plastic bags. Figure 2-3 illustrates the impact of 1 million paper vs. plastic bags on the environment. Table 2-2 provides a summary of the data collected by the students in one of the studies. Before the study, the students did not comprehend the amount of energy and chemicals used in the processing of paper. The

Table 2-1 Environmental Impacts of Manufacture of 1 Dry Ton of Wood Pulp

Item	Kraft Unbleached	Kraft Bleached	Groundwood
Virgin material (dry, tons)	1.0	1.1	1.1
Wastewater volume (gal)	24×10^3	47×10^3	10×10^3
Energy (BTU)	16×10^6	22×10^6	18×10^6
Solid wastes (lb)	135	225	164
Air emissions (lb)	50	64	65
Waterborne wastes[a] (lb)	46	90	25

[a] Includes BOD.

energy and chemicals involved with recycle of paper bags was seen as something most people would not have considered. The quantities and disposal of the wastes from the processing was a shock to the engineering student. There is no reason to think that a similar study of industrial products would not open the eyes of many senior managers.

Another example is a manufacturer of computer terminals. In this case the study started with the component manufacturing, i.e., semiconductor chips, semiconductor packages, printed wire board fabrication, displays, computer systems and assembly, use, and disposal (Figure 2-4). In each step of the processing, a study was made of all the material and energy inputs and outputs. This was a very large study. The report took multi-three-ring binders. The

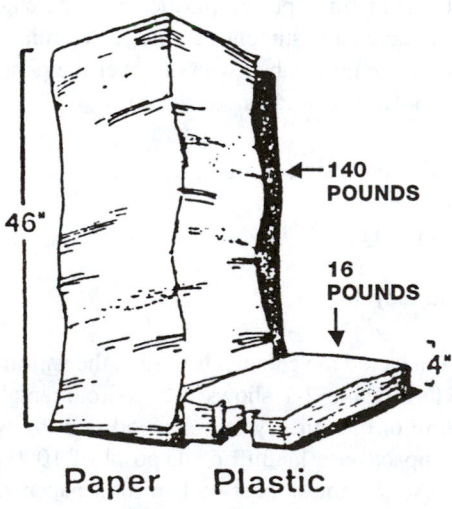

Figure 2-2.   **Impact on a landfill of disposal of 1000 paper grocery bags vs. 1000 plastic grocery bags.**

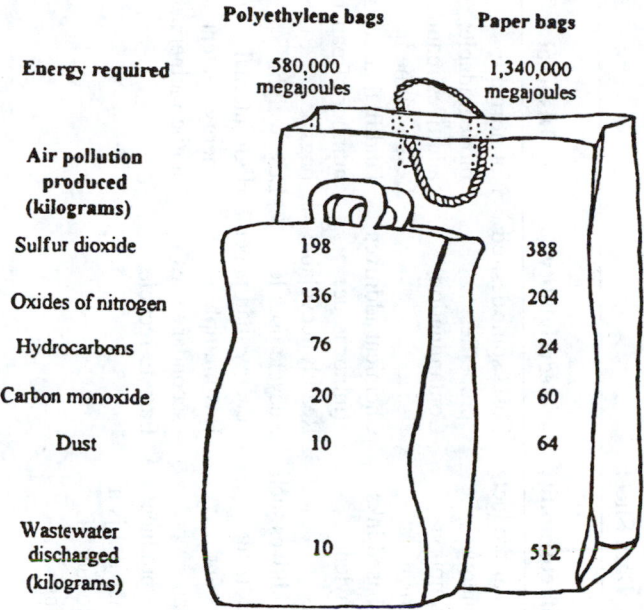

	Polyethylene bags	Paper bags
Energy required	580,000 megajoules	1,340,000 megajoules
Air pollution produced (kilograms)		
Sulfur dioxide	198	388
Oxides of nitrogen	136	204
Hydrocarbons	76	24
Carbon monoxide	20	60
Dust	10	64
Wastewater discharged (kilograms)	10	512

Figure 2-3.   The energy dilemma. The environmental burdern per 1 million bags.

report resulted in reductions in chemicals used, changes to chemicals used, and lower energy consumption. Examples of chemicals originally used include

- Sulfuric acid
- Phosphine
- Xylene
- Glycol ethers
- Gold
- Isopropyl alcohol
- Epoxies
- Plastic molding compounds
- Polyethylene glycol
- RMA flux
- Alloy 42
- Nickel salts
- Metal resists
- Terpenes
- Chromic acid
- Yttrium compounds
- Hydrochloric acid
- Arsine

Table 2-2 Paper vs. Plastic Data Sheet

Process	Energy	Environmental impact	Manufacturability	Recyclability	Disposal
Plastic	580,000 megajoules per bag required Can be incinerated to produce energy High internal energy content (BTU/lb) Uses about half the energy it takes to produce an equal number of paper bags	Thermal decomposition produces aldehydes Water wastes produced through production Littering of postconsumer plastic; trash on streets, oceans, and wilderness areas Primary source of raw material is nonrenewable Can use renewable sources but more expensive	Inert final product Low energy consumption Gas blown Water-based inks Polyethylene pellets Uses additives such as oxidizing agents and colorants Easy to produce Raw material is a by-product of refining	Loses some desired properties Contamination issues Chemical additives (minor issue) Readily recyclable Recycle uses less energy than paper Has multiple secondary uses Easy to recycle	Not normally biodegradable; additives required Less volume in landfill than equal number of paper bags Less weight High internal energy content Can be incinerated

Paper	1,340,000 megajoule per bag required	Odor producing	Messy manufacturing	Low profit margin in recycling	Increases solid wastes
	Numerous equipment transfers required to complete a finished product	Produces air and water pollutants	Numerous operations to finish the product	Limited degree of reprocessing	Not readily biodegradable
		Inorganic fillers such as ink and adhesives produce large quantities of wastes	Government rules have increased costs	Secondary fibers are lower quality	Large relative volume
		Deforestation		Very high capital investment	
		Carcinogenic byproducts produced			
		Not readily biodegradable			

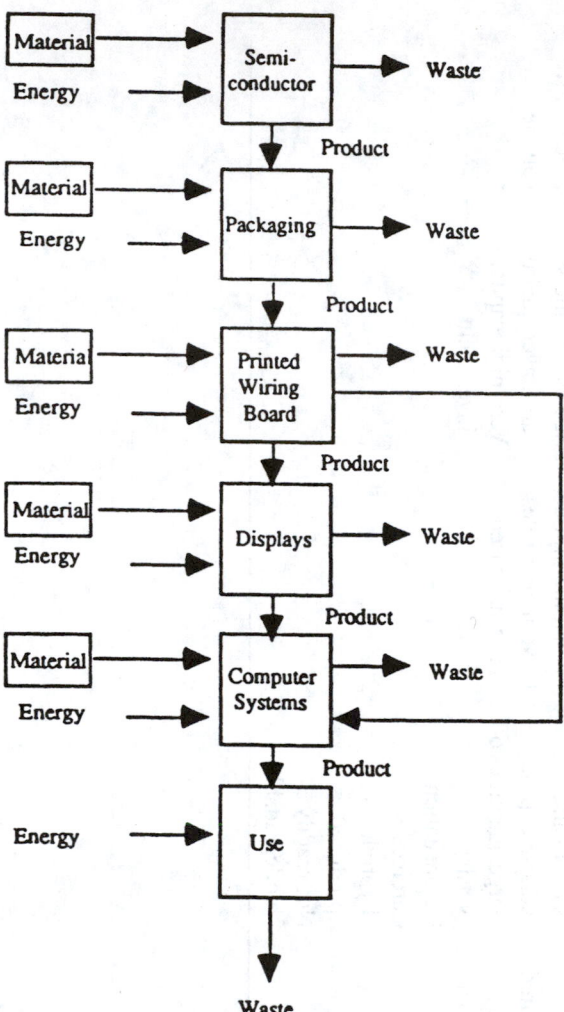

Figure 2-4A.   Top level schematic for the manufacture of a computer.

- Acetone
- Diborane
- Cyanide
- Aluminum
- RTV
- Dioctyl phthalate plasticizer
- Tin/lead solder
- Copper compounds
- Sodium hydroxide

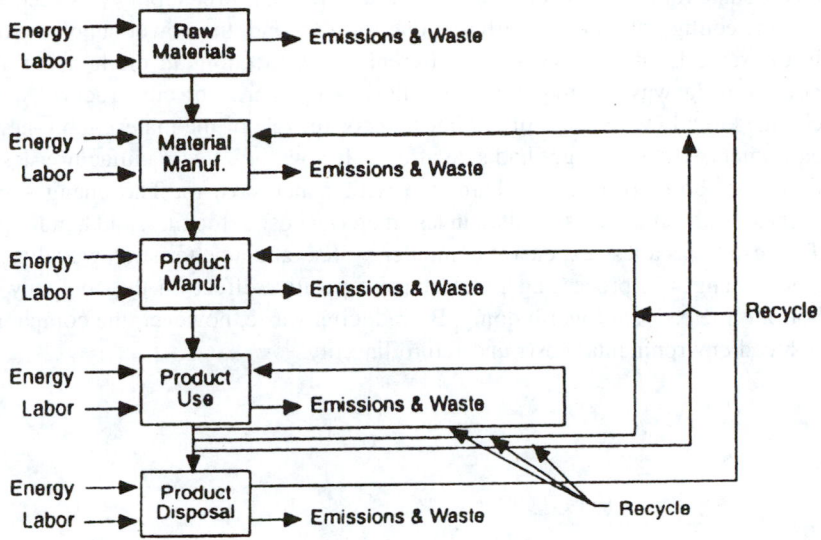

Figure 2-4B.   Basic product flow diagram. (From *Waste Minimization as a Strategic Weapon,* **Ciambrone, D. F., CRC/Lewis Publishers, Boca Raton, FL, 1995.)**

- Fluoboric acid
- Organic solvents
- Zinc sulfate
- Urethanes

and many others. One can readily see the opportunities for possible savings. This list does not include the use of energy. During the manufacture of the semiconductor chips, energy is used to power evaporators, ion diffusion equipment, and etchers. Plating baths use large quantities of power. For example, when plating printed wire boards with copper, the plating bath required 25 Amps per square foot of area being plated. An 18" × 24" panel, plated on both sides, requires 150 A. If there are eight panels in the tank at a time, this will require 1200 A. The injection molding equipment used to make the plastic housings, keyboard, etc., requires large electric motors. Machining the dies used on the molding equipment requires large motors as well. All of these processes have waste. This waste must be disposed of. Some of the waste or by-products are vapors and fugitive emissions that affect the environment. Reducing/changing the materials and processes can result in savings to the company and the environment.

Some of the savings mentioned resulted by redesign of the product. For example, by redesigning the computer keyboard, multiconfigurations could be built on the same basic base and subcomponents. The display housing would be the same for different configurations. If a different power supply was needed for one configuration and a different one for a second, the power supply went in the same location, fastened to different screw locations in the housing. A new housing was not required. The choice of material of construction was changed to allow for reuse of old housings or recycle of the material into new equipment. These changes had side affects. It now took the manufacturer less time and labor to assemble hardware with a net savings. The changes in processes and chemicals resulted in less material costs, stocking, and handling. The result was a less-expensive computer system and a better strategic advantage. Changes in processing should have a positive effect on the company's bottom line or it isn't worth doing. By reducing waste, however, the company reduced environmental costs and future liability.

3   LIFE CYCLE FRAMEWORK

SCOPE

This chapter discusses the procedural framework for performing a life cycle inventory. While there is agreement on the major elements of an inventory, procedural discussions occur at many phases of the process. To perform a study whose intended use is to evaluate the environmental burden (MANPRINT) associated with a product, process, or activity, the scope of the study must be clearly defined. The following components, at a minimum, should be defined:

- How the results will be used
- The product, process, and activity to be studied
- Reasons for conducting the study
- The elements not to be addressed
- The elements of the analysis

At the onset of the study, the level of detail must be decided. Sometimes it is obvious from the application or intended use. In other circumstances, there will be several options to choose. These will range from generic study to one that is very product or process specific. Most life cycle studies fall between these two extremes. Even at the product-specific study level the company can determine whether their need can be met by one of the following approaches:

- Full scale life cycle
- Partial life cycle
- Individual stages or processes

The partial life cycle approach consists of several stages. This allows for stages to be omitted if they have been studied before and are static for purposes of this study. The life cycle analysis can also be limited to specific stages or processes. This would be the case if a new process is being considered or if the product finds another use or market.

To assist in providing focus and direction, it is extremely important to clearly understand how the results will be used. Management needs to be able to clearly define why the study is being undertaken, who will use the data and why. The potential users need to understand what is being supplied and what they are expected to do with the information. This will help define how the data are taken and formatted for future use.

Probably the most enterprising step is to define what is to be studied. The organization has a reason for wanting to conduct the study. Now the issue is exactly what to study. This sounds ridiculous but, unfortunately, it's true. Here we must not only specifically define which product or process to study but which aspects of it. In some cases, companies have tried to study the "whole" life cycle and it wasn't too difficult. Their product was early in the "food chain." A logging company looking at their impact on the process of logging wouldn't have as large a study as a company making airplanes. Obviously, a study could be made on airplanes, assuming someone wanted to pay for it. A supermarket studying paper or plastic bags would not be focusing on vegetable deliveries or their containers.

The depth of the study is linked to the reason for wanting it. In some cases, certain aspects will be studied in depth and other areas in less depth. It is not unusual to look at certain selected processes or define the major areas of the life cycle of interest and use the data to compare alternative processes or products. In the case of airplanes, a company wanted to look at after-sale service and repair. This was to determine the impacts of the as-designed plane and what improvements could be made in the aircraft to help minimize these impacts. For example, a part of the study found that, during customer maintenance, stripping paint from the planes and repainting created a large amount of hazardous waste. The waste included paint strippers and the dissolved paint. The local air quality management authority would be interested in the impact on air. The airplane manufacturer had the same problem to a lesser degree. The study of the maintenance cycle found that heavy abrasives damaged the aluminum skin of the planes. Sandblasting could also cause damage, depending on the parameters of the setup and where it was used. They found that by blasting with very small carbon dioxide pellets the paint came off with no damage to the airplane's skin. The process did not create air pollution and the only residual was the solid paint. The carbon dioxide sublimated to a gas leaving just the solid paint. The solid paint was not considered a hazardous waste. This saved not just all the paperwork, health, safety, and environmental issues but also the costs of disposing of a hazardous waste.

A summary study could also be made to provide management with areas that would potentially lead to cost or environmental improvements. This approach would provide for future focused studies.

As indicated above, knowing what to study is half the problem. The other half is to define what is *not to be studied.* For example, " This study will not address aesthetic and socioeconomic issues." By understanding the purpose of the study and the system boundaries (what is and isn't included), it will assure a valid interpretation of the results.

To follow the process one needs to understand some basic definitions that will be used in the text and graphics. The following short definitions will provide the reader with the knowledge required.

In life cycle analysis, the "system" refers to a collection of operations that together perform some well-defined function.

Life cycle inventory accounts for every *significant* step in a product system. To assist, system flow diagrams are used. The diagrams assist with necessary calculations.

DEFINING BOUNDARIES

Boundary conditions refer to what is included and excluded from each step in the process. Since various aspects may be bundled, the boundary condition defines what is bundled and not treated independently. For example, a study might include transportation, distribution, and storage as one "operation" because other aspects of the product life cycle are more important at the time.

A complete life cycle inventory includes and quantifies resource and energy use and environmental releases throughout the product life cycle as illustrated in Figure 3-1. The major stages considered include the following activities:

Raw materials acquisition and energy. The boundaries of this element include all activities needed for the acquisition of all raw materials

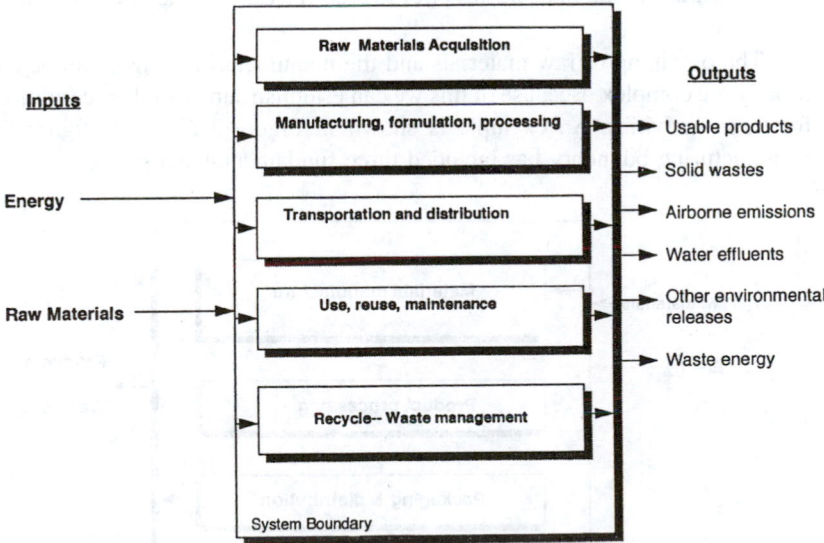

Figure 3-1.   Life cycle inventory.

and or ends with the first manufacturing or processing stage that refines the raw materials.

Manufacturing, formulation, and processing. The processing step takes the raw materials and converts them into products.

Transportation and distribution. This aspect of a life cycle analysis is common to almost all such studies. In this stage, the product undergoes a physical change in location or configuration. It does not involve a transformation of materials. Transportation is movement of materials and energy involved. The transportation and distribution stage might involve multiple activities at various times in the life cycle. Transportation during the raw materials and manufacturing stage is usually included in the these stages. The stage for transportation and distribution involves those activities used after manufacturing.

Use, reuse, and maintenance. The boundary condition for this stage begins after the distribution of product or materials and ends at the point where the product or materials are discarded and enter a waste management system. The impact the product has on the environment during use/reuse will vary with each product. If the product uses energy, fuel, or another natural resource, this is an important stage.

Recycle, waste management. The recycle step involves reclaiming materials out of the waste management system and returning them to the manufacturing or processing stage. Waste management is the effective disposal of any material released to any environment — land, air, or water. Waste streams are generated at each phase of the life cycle. Waste management includes any mechanisms for treating, handling, or transportation of wastes prior to their release into the environment.

The obtaining of raw materials and the manufacturing stages can be and usually are complex. Because of this we can establish sub-boundary conditions for these activities. An example is shown in Figure 3-2. In the figure the manufacturing boundary has included three fundamental steps.

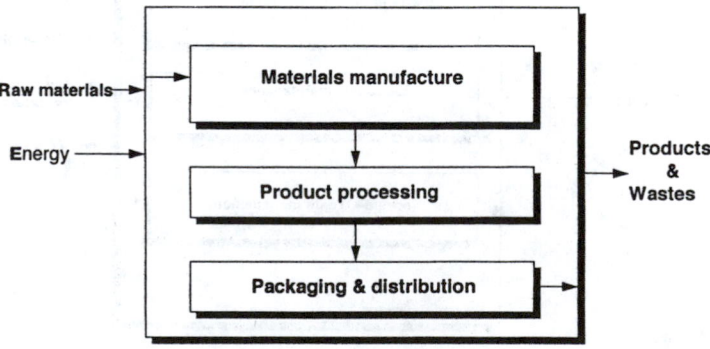

Figure 3-2.   Manufacturing stage.

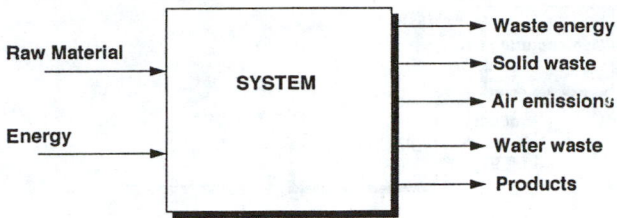

Figure 3-3.   Top layer model.

To keep the analysis under control, the mapping of the "processes" to be included needs to be done. This will enable the process to be visualized and facilitate the proper setting of boundaries and depth of analysis. This can be accomplished using a simple method of modeling. In this "systems" model you will start at the *top* layer as shown in Figure 3-3. In Figure 3-3, the whole of what is to be studied is in the box. This is level 1.0. Level 1.0 can be broken down into the next layer. For example, the owner of "Acme Plastics & Plating" does injection molding of plastic parts and chrome plates them. He is interested in the life cycle impact of the plated plastic shower heads on the environment. To do this he will model the life cycle of his product and decide the boundary conditions based on his model. The owner has decided to exclude the processes of making the ABS plastic and the manufacture of the plating solutions he purchases. These will be considered raw materials to him. The top level system diagram (1.0), therefore, is shown in Figure 3-4.

Level 2.0 further breaks the process down as shown in Figure 3-5. Each block on the model has a number assigned based on where it came from. For example, in the level 2.0 model layer, each block has an identification number, such as 2.1 for manufacturing. This allows the user to take each block and systematically break it down as far as he wants to without losing track of where it belongs in the system. This is further illustrated in Figure 3-6. In this figure, block 2.1, manufacturing, was expanded. We further expanded this in Figure 3-7 by breaking down block 2.1.3. Each of the blocks in Figure 3-6

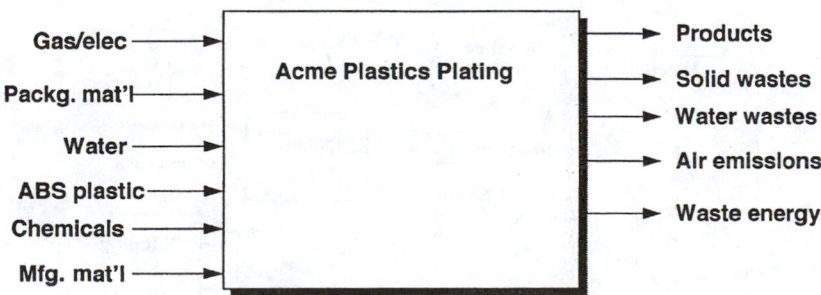

Figure 3-4.   Acme level 1.0.

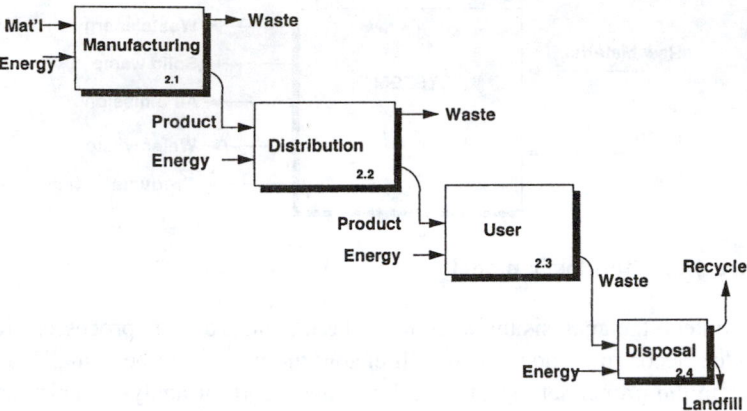

Figure 3-5.   Level 2.0 model.

can and should be further broken down. As a rule, each breakdown should not have more than four to six blocks. In our example, Figure 3-7 is as far as we need to go for this block. Modeling in this fashion provides the following advantages:

- Understand the process flow in detail
- Simple means to verify the flow
- Identify and quantify the wastes and products from each step
- Insure you captured all the required information
- Facilitates trying a "what if" scenario process

The models can actually have quantifiable numbers written in for each arrow on the incoming and outgoing side of each block or allow the use of formal check sheets for each block. If the quantifiable numbers for such things

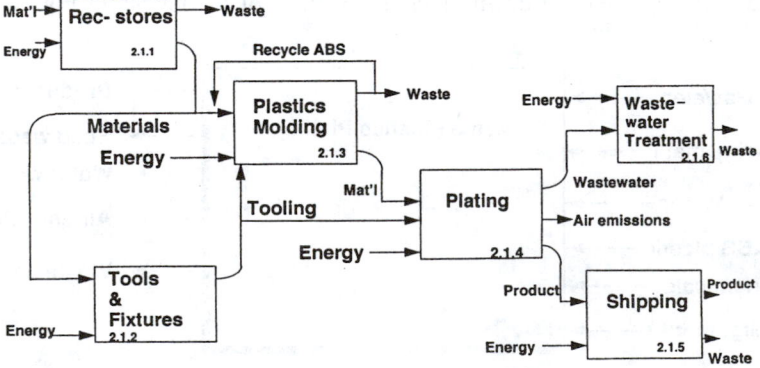

Figure 3-6.   Manufacturing.

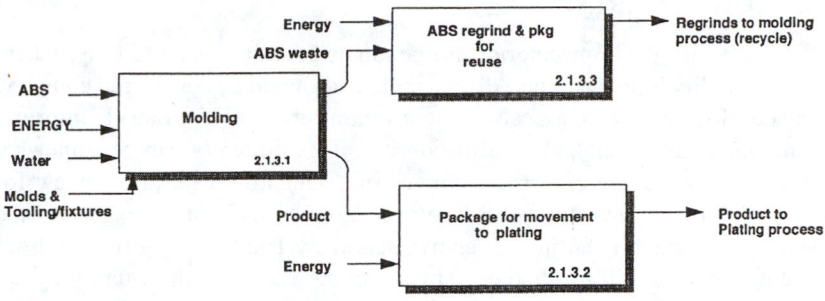

Figure 3-7. Plastics molding.

as energy is not available at each step, the analyst may try to arrange for measurements, calculate the energy based on data about the equipment (HP of motors, etc., for the time of analysis), or find the level where it is available and partition it as best as possible. For example, the user could take the total gas and electric usage for the facility and apportion the approximate amount to the areas under study. Whatever method is used, make special note of it for future reference. Whenever possible, the data describing the performance of the subsystems should be drawn from primary sources. In some cases, the manufacturing subsystem will be unique to the system being studied. When secondary or published data sources are used, the data source should be clearly referenced. Any inherent limitations in the data should be clearly identified.

When conducting the study, it is essential that the operating period chosen be long enough to smooth out any atypical behavior such as machine downtime, setup operations, or changes in flow due to stock/material fluctuations. A period of 9 to 12 months is found to be best. Although it is tempting to use industry averages, it is better to use actual data from the processes being studied whenever possible.

CHECKLISTS AND THEIR APPLICATION AND USE

The use of this type of modeling facilitates both simple and complex processes and systems. The model makes use of data gathering and check sheets for easier understanding and minimizes missed data. A sample check sheet is shown in Figure 3-8. Because the user may want to determine where the MANPRINTS are the largest or where there are opportunities in the product stream to minimize environmental impacts, this method can provide a less-costly means of analysis and data gathering than conventional methods.

When the user actually models the product flow it will become evident that certain processes are interdependent on others. Therefore, a change to one area needs to be analyzed to determine the impact, if any, downstream. In the case of Acme Plastics & Plating, if the engineers changed thermoplastic to improve the molding process and reduce waste, there might be a major impact (positive or negative) on the plating operation.

DATA GATHERING

Many life cycle inventories are conducted for internal use only, where confidentiality is not an issue. All data collected should be considered company private. However, some aspects of the inventory may require outside information. Obtaining accurate data is of concern. This difficulty can be somewhat overcome by the use of outside sources of "standard" data, peer review for external data and inventories be developed to provide for maintenance of the generic information and insure relative accuracy. The acquiring of data from external sources will to some extent rely on the use of confidentiality agreements. There are many sources of public information for use in the life cycle inventory. Examples of these include:

- U.S. Department of Energy (DOE) databases of energy uses by industry aggregate
- U.S. Environmental Protection Agency (EPA) databases include the range of chemical quantities used by site and emissions estimations
- Local/regional gas and electric companies
- Open literature, including journals, patents, trade publications
- Trade/technical associations publications and databases
- U.S. Department of Commerce (DOC) databases
- U.S. Department of Defense (DOD) databases

The use of case-specific and industry average data is an important issue. No requirements exist that define what type of data is to be used for the analysis. The quality of the data can have a drastic affect on the life cycle inventory results. Because of this, it is important to be aware of the source of the data, the type of data (industry average, literature data, specific process data, etc.), worst case or average or best case, and the limitations. Data on specific environmental emissions are preferable in the inventory whenever possible. The data should also be normalized for the actual amount of product being produced, not the plant capacity. There will be sections of the life cycle analysis that can and should use industry data to control cost of the analysis. In all cases, the most recent data should always be used.

When evaluating wastes from any process, all the wastes must be accounted for. Wastes from some processes are easy to quantify. They are solids and can be weighed. In the cases of the injection molding process for Acme Plastics & Plating, the wastes are plastic (weighable), water from the molding die cooling line (measurable — gal/min, heat — temp), and fugitive heat (nonmeasurable to any accuracy). In other cases there are mainstream and fugitive emissions. Both should be used. When an emission is not measurable, a qualitative discussion is required, and conclusions should explicitly reflect this deficiency or condition. There is another release that needs to be considered in the time period being studied: the accidental release. The inclusion of accidental releases should be made based on history of the process. It can be included as a correction factor and added to routine releases.

During the data gathering, it is likely that you will encounter gaps in the data. This will be because the information is not available, is inconsistent, or is reported as nondetectable. It is important to note what data were not available and why. In the case of the not available category, information is sometimes lumped in with other data that the accounting or management uses. Sometimes it is extractable. However, it usually isn't worth the effort or cost to obtain it. In some instances, data are available from the people working on the processes being studied. Their data are usually more accurate and up to date. Some information that management thinks is unavailable is obtainable from the workers in the area in question. Don't forget to talk to the people actually doing the work. Sensitivity analysis will be covered in a later chapter.

CONSTRUCTION OF A CONCEPTUAL MODEL

To construct the basic models the user will make a "flow diagram" of the overall product/process flow to be considered in the study. This can be along the lines of that shown in Figures 3-5 and 3-6. Next the user will make detailed process diagrams as shown in Figure 3-7. If more detail is needed then further refinements at lower levels will need to be made. Consult with process/product experts to insure model and data requirements accuracy. Once the auditor is satisfied that he or she has enough detail in the models to satisfy the audit requirements, then the data gathering will begin. This way the auditor knows exactly what data are required. This saves time collecting useless information. The sample checksheet as shown in Figure 3-8 may be used for each process for information gathering. Next the auditor will assemble the data on each applicable diagram to characterize each individual process (see Figure 3-9). This process allows for the accounting of all the information and identifications of data gaps. The diagrams can now be put together and the data added at each process detail level for inclusion on the next level diagram above them. For example, if Figure 3-9 is one of the elements on Figure 3-7, and all the other "processes" on Figure 3-7 are complete, then their totals would be entered on Figure 3-6 for process block 2.1.3 Likewise, all the data on Figure 3-6 could be totaled and entered in Figure 3-5 for block 2.1, etc.

Starting at the top and breaking everything down to the lowest appropriate level allows for an understanding of the total operation and prevents omissions. By starting at the lowest level and researching/entering the required data, it allows for a systems accounting of all the data and identification of gaps. Spreadsheets can sometimes help (see example in Figure 3-10). This upward linking prevents inadvertent omissions and double counting errors. During this operation the use of standard metrics will be applied. This includes the conversion of energy fuel value into standard energy units, such as million BTU or gigajoule. This principle also allows the user to identify possible reuse or

ELCA DATA CHECKSHEET		
ELCA PROJECT NAME:		
INVESTIGATOR:		
DATE:		
MODEL SUBASSEMBLY BLOCK REFERENCE NUMBER:		
PROJECT ID NUMBER:		
COMPANY		

INPUT ITEMS	QTY.	UNITS
RAW MATERIALS		Kg/lbs/gal/ltr/cuft
Energy		KW/Hp/MJ
OTHER		
OUTPUT ITEMS		
RAW MATERIALS		Kg/lbs/gal/ltr/cuft
ENERGY		KW/Hp/MJ
OTHER		

Figure 3-8.   Sample checksheet.

recycle applications for wastes from each process. For example, the waste hot water in Figure 3-9 may be able to be used as a final rinse in the plating operation at the plant or to heat a plating tank. This would allow for less energy costs to the plant and dual use for the water. The discharge may be usable for irrigation or other use if not contaminated. One will notice the impact of one area or process on another. It is imperative that the auditor keep track of the interdependencies between processes. It should be clear that an action in the manufacturing can impact something in distribution, use, or disposal.

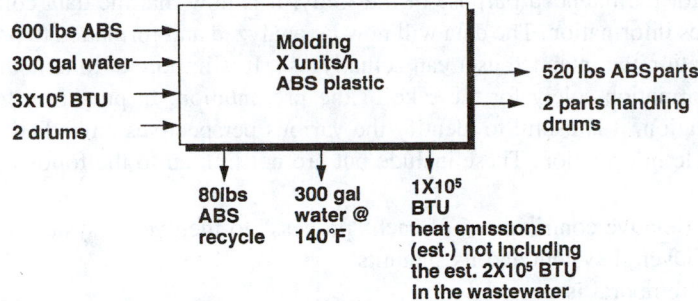

Figure 3-9. Data model example. There are six identical molding machines, each running two 8-hour shifts per 5-day week.

INTERPRETATION AND REPORTING OF RESULTS

When writing a report to present the final results of the life cycle inventory, it is important to consider the following:

- The methodology used
- The *intended use* of the study
- The boundary conditions set
- All assumptions used
- The *intended users* of the study

Life cycle inventories generate a great deal of information. The auditor/analyst needs to select the presentation format and content that are best

Purpose: What is being studied ? Why ?								
Scope / Limitations: What are the boundaries ? Assumptions ?								
	Inputs				Outputs				Other
Process number	Air	Water	Chem./Mat'l	Energy	Air	Water	Chem./Mat'l	Energy	
No.								
Raw materials								
Materials								
Packaging								
Products								
Wastes								
Energy								

Use spreadsheet for each process under study.

Figure 3-10. Sample spreadsheet for data integration.

suited for the intended purpose of the study. It is now that the data collected becomes information. The data will now be analyzed and formatted to provide information the intended user can actually use. It is important to not simplify the information solely for the sake of the presentation. In preparing for the presentation, it is useful to identify the various perspectives embodied in the life cycle information. These include but are not limited to the following:

- Relative contributions of each "process" to the overall system
- Overall system and its subunits
- Temporal issues
- Data parameters groups within each category, e.g., water wastes, air emissions, solid wastes
- Data categories across processes
- Geographic conditions relevant to the study
- Data parameters within groups

The analyst must present the information in a manner that increases the comprehension of the results without oversimplifying them. It is obvious, therefore, that knowing who is going to use the results and for what purpose will simplify the analyst's job of presenting the information. The information can be presented in graphic and tabular formats. The author suggests a combination of both. The following suggestions should be considered when presenting the information.

- Provide total energy results
- Provide energy results broken down by "process" (process is each stage or useful subgroup)
- Present wastes as industrial, package/distribution, consumer, and postconsumer
- Categorize air emissions, waterborne wastes, and industrial solid wastes by process
- Present energy utilization by major category or by process

The use of tabular presentation is very comprehensive. The choice of how the tables are formatted varies based on the scope and use of the study. If the study was to determine what packaging system to use, the overall system results can be used. If it is to compare two competing packaging systems, again the overall system results would be of interest. However, if the study was performed to determine how the packaging could be changed to minimize the impact from its use on the environment, then it is important to not just present the overall results but contributions made by each component of the packaging system. For example, in the case of the plated plastic parts from Acme, it would be necessary to consider the plastic bag each part is placed in, the cardboard box it is placed in, the corrugated box the individual boxes are packaged in, and the stretch wrap around the boxes. If a pallet is used as part of the shipping then it will be included. The user can concentrate improvement efforts on the components that make substantial contributions to minimizing the environmental impact of the packaging.

Graphical presentation of information helps to augment tabular data and aids comprehension. Pie or bar charts or graphics such as drawings, flow charts, photos, or combinations are useful. These types of charts help the user take ownership of the information. When making the report and charts, the analyst needs to ask and answer questions about what each graph is intended to convey. It may be necessary to use multiple charts to convey the intended message. Figures 2-2, 2-3, 10-4, 10-5, 10-8, and Appendix C illustrate the use of multiple format graphics.

As noted earlier, it is imperative that the analyst understand the intended use of the study. It will facilitate determining what type of data is most applicable, for example, industry wide or mostly site specific. It will assist in the preparation and the presenting of the information. Understanding the user and the intended use of the study will prevent misunderstandings and unintended use of the study. The effective communication of the results must include an understanding of the assumptions made and the boundary conditions placed on the study. Careful interpretation is required to avoid making unsupported claims.

An important aspect of the interpretation or understanding of the study is the data accuracy. If the data are "industry wide", the user must be careful in their use. The same process conducted at different plants may use similar but different materials, technologies, equipment, and be of different age and efficiency. The different plant sites may be operating under different government regulations such as environmental. Factors such as these need to be understood before using the industry data. If the data are from industry averages and not heavily weighted in the overall study, use may contribute to the conclusion but, in itself, is not a driving factor. These data are subject to both systematic and random error. Site-specific data are true random error. This type of error can be described in conventional statistical terms using mean and standard deviation of the measured results. However, systematic variation due to feedstock or technology/equipment differences is not random error. These sources of variability are "explainable" by virtue of the conditions, technology, age, or other identifiable factors of the plants used. The analyst should explain the importance of these sources of variability to the user. For example, some plating equipment uses steam for the source of heat while others use electricity.

Boundary conditions and data for many internal life cycle studies may require the interpretation of the results for a particular corporation. The results may be very specific to that particular company. This would preclude the use of industry standards or averages. Therefore, a high degree of accuracy may be assumed.

Most life cycle studies are for the internal use of the company or organization that chartered and funded it. When results of life cycle analysis are released to the public, however, they are usually used to support a specific statement by the company. It is important that the information provided was not selectively derived and that all pertinent information is included

The use of published studies comparing competitive products, processes, materials, or practices must be presented cautiously. The assumptions, boundaries, and data quality should be considered in presenting conclusions. Studies with different boundary conditions may have different results but both may be accurate.

4   GENERAL ISSUES

Before a life cycle inventory is started, it is important that each of the participants understand their respective role. The following summarizes the main role of key participants.

Life cycle participant	Responsibilities
Accounting	Assign environmental costs to products, processes, calculate hidden liabilities, etc.
Marketing/advertising	Give feedback on existing products and demands for alternatives, value of low environmental impact, inform customers of reduced environmental risk and benefits
Distribution/packaging	Provide data requested; provide reduced impact concepts and estimate of cost/environmental impacts-risks
Government regulators Standards organizations	Provide copies of rules/regulations/laws, provide copies of studies/reports, provide environmental impact data previously developed
Design engineers	Provide information on materials and "processes" used in present products and what is being considered on new products; provide data on concepts for reducing environmental impact and any cost studies associated with the concepts
Manufacturing engineers	Provide data on manufacturing processes including inputs and outputs, wastes and pollution prevention activities in place or underway; materials and process inputs and outputs
Management	Establish corporate environmental policy; charter the life cycle inventory and analysis, insure follow up and implementation of proper and cost-effective activities arising from the analysis

Life cycle participant	Responsibilities
Purchasing	Provide the input on products/raw materials including specifications, costs, and known environmental impacts; assist suppliers in reducing environmental impact by sharing company data cleared for release by management
Waste management professionals	Offer information on the fate of industrial wastes and retired consumer products; environmental impacts and risks, alternative waste disposal methods and costs and risks associated with the alternatives, information on recyclability and markets
Service	Provide input on repair and maintenance processes and concepts for improvements

TEMPLATES

During a life cycle analysis there will be issues that are encountered in evaluating and using the information. Most references emphasize the use of templates. The key issues to be discussed in this chapter include

- The use of templates
- Boundary conditions
- Data issues and concerns

A life cycle analysis can become quite complex in a very short time. As illustrated in Chapter 3, the easiest way to map out the product flow and processes involved is to use diagrams or templates. A template is a guide used to aid the analyst in collecting and allocating data. Product flows can start as simply as the one shown in Figures 2-4 and 3-5 and the Appendix. However, it becomes obvious that this level of detail is inadequate. Detail diagrams are required. More detailed process flow charts for the "manufacturing" processes are illustrated in Figures 4-1 and 4-2. Examples are illustrated in Chapter 3. Some texts use the basic template shown in Figure 4-3. The basic templates used must be developed to the lowest level (subsections) necessary to obtain the required data. The diagrams assist in keeping track of the information. Whether the analyst uses the contemporary format, Figure 4-3, or the type illustrated in Chapter 3, they all attempt to do the same thing. Each chart will have

Inputs	Outputs
Raw or intermediate materials	Air emissions
Energy	Waterborne wastes

Inputs	Outputs
Water	Solid wastes
Other inputs	Products

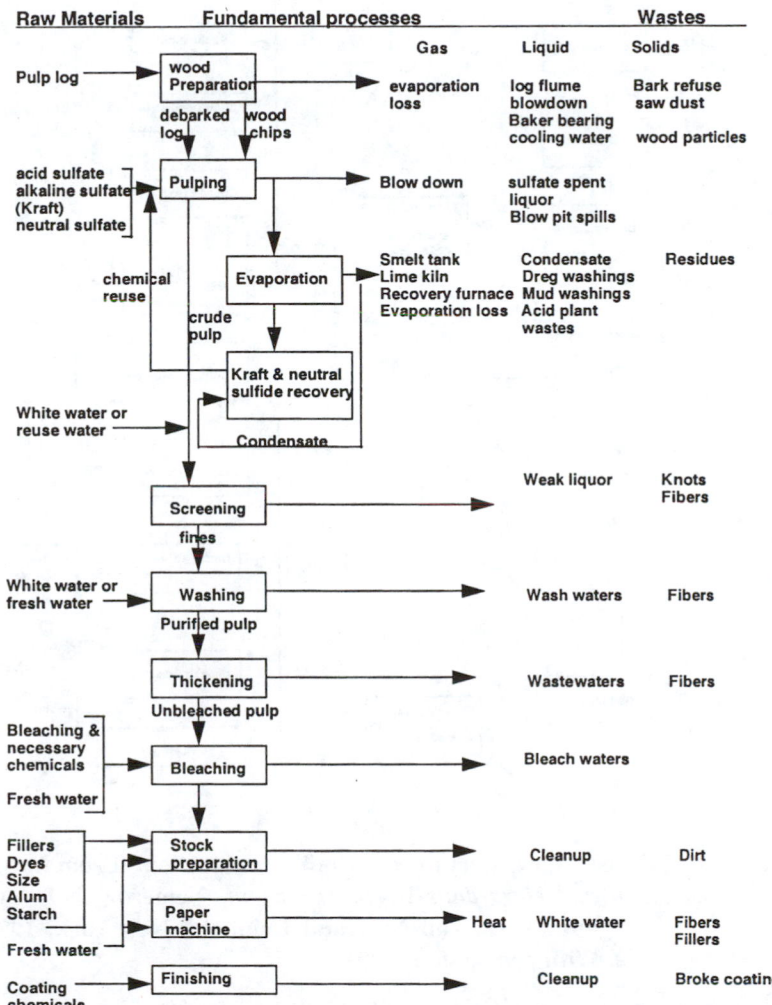

Figure 4-1. Simplified diagram of fundamental pulp and paper processes. (From *Industrial and Hazardous Waste Treatment*, Nemerow, N.L. and Dasgupta, A.D., Van Nostrand Reinhold, New York, 1991, 928. With permission.)

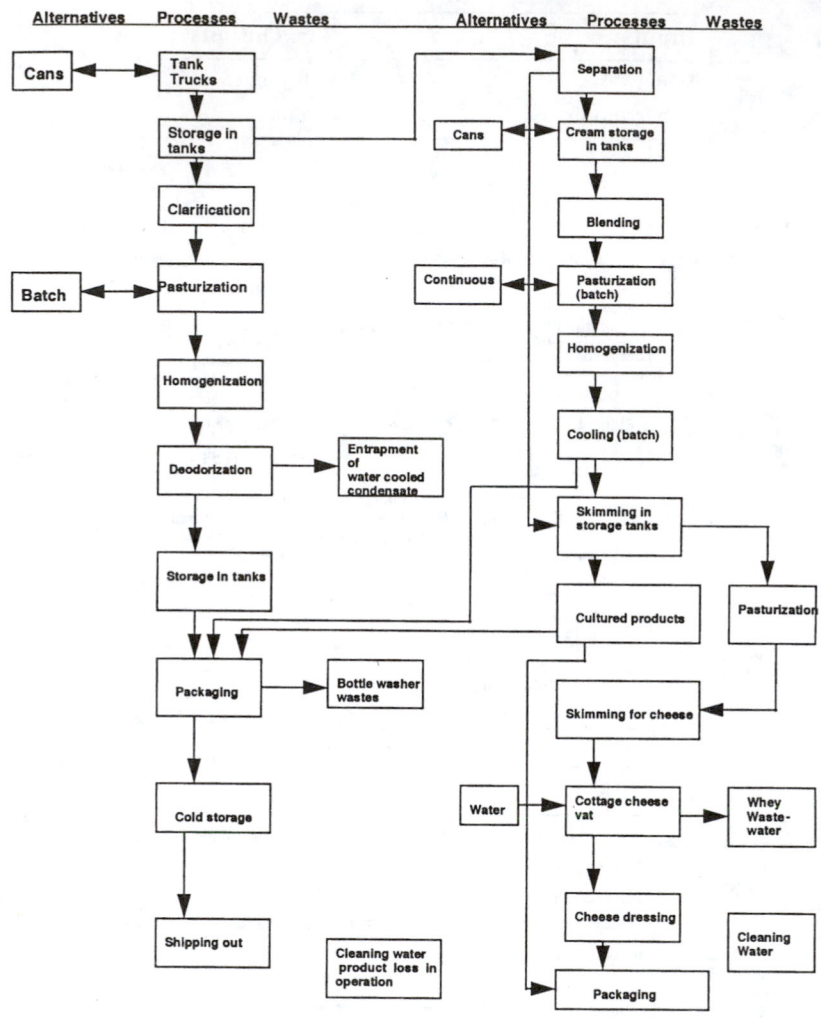

Figure 4-2. Process flow chart for fluid-milk preparation. (From *Industrial and Hazardous Waste Treatment,* Nemerow, N.L. and Dasgupta, A.D., Van Nostrand Reinhold, New York, 1991, 462. With permission.)

DATA ISSUES

The thing to remember is that in some cases the materials input is actually the product of a previous step. With that product may be packaging materials that will become waste or raw materials. During the analysis, any waste

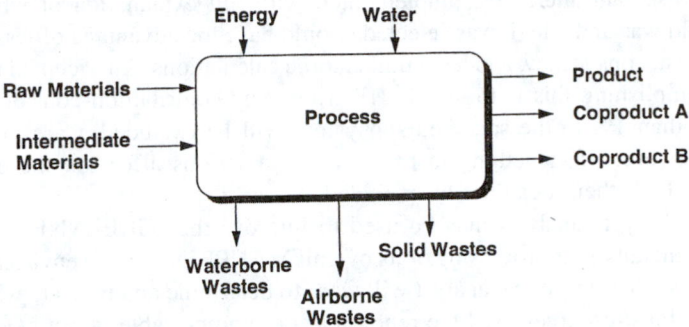

Figure 4-3.   Life cycle inventory template.

treatment system used during a process is considered a process and should be modeled as such.

The input materials for each subsection are referred to as raw or intermediate materials. Raw materials are usually materials that have been extracted from the earth but have not been refined or manufactured. Examples include mineral ore, animals, plant products, etc. Intermediate materials are those that have undergone some level of processing. The most complete life cycle inventory will begin at the raw materials level. In the case of the milk in Figure 4-2, it may be considered an intermediate because of the cows. There is an impact on the environment from the raising and keeping of cows. Others may consider the milk a raw material since nothing has happened to it prior to arriving at the milk processing plant. If one considers the milk to be the starting point, then the rationale for the decision must be noted. The same rationale may be applied to the pulp/paper process in Figure 4-1.

Deciding which raw/intermediate materials to include in a life cycle inventory is complex. Several options are available:

- Include all materials no matter how minor.
- Within the scope of the study and boundary conditions set, exclude inputs less than a predetermined and clearly stated threshold.
- Within the boundary and scope of the study, exclude inputs determined to be negligible relative to the intended use of the analysis.
- Within the scope of the study, consistently exclude certain classes or types of inputs. Again, clearly define this exclusion and why the decision was made.

The advantage of the first approach is that the analyst does not have to explain or defend what has been included or excluded. The disadvantage is that the whole study could become a long, expensive exercise. The computational aspect of including everything could be exhaustive.

The second alternative, implemented with full explanation of what the threshold was and why it was selected, would have the advantage of less time, lower cost, consistency, and less cumbersome calculations. An accepted means of accomplishing this is to use the 1% rule. Any contribution considered to be less than 1% to the specified subsystem will be excluded. Care must be used, however. If something in that 1% has a serious affect on the overall MANPRINT, then it can not be excluded.

A life cycle analysis may be used to improve the "GREENNESS" of a product including its processing or its overall MANPRINT on the environment. In cases such as these, the analyst will want to determine and indicate whether the materials/resources used are renewable or nonrenewable. A set of definitions for these two terms are

- Renewable material/resource — A renewable material or resource is one that is being replaced in the environment in a time frame relevant to society.

Examples include certain species of wood, farm crops, kelp, etc. Note that the materials/resources must be renewable and be replenished in a time frame relevant to society. Some materials that would have been renewable have been so overused that they are no longer considered renewable.

- Nonrenewable material/resource — Nonrenewable materials/resource is one that is *not* being replaced in the environment in a time frame relevant to society.

Examples include minerals mined from the land, coal, oil, natural gas, and materials from exhausted supplies.

Energy is shown as an input on templates. The energy shown at higher levels is a composite of those at each subsequent subsystems. There are three classes of energy for use in the analysis: process energy, transportation energy, and energy of material resources.

Process energy is required to operate and run each process in the subsystems. These include electrical power for motors, pumps, reactors, plating cells, etc. and gas for heating. Transportation energy is the energy required for various modes of transportation such as trucks, ships, airplanes, rail carriers, barges, and pipelines. Forklifts and conveyors could be considered either as transportation or as process according to their role in the subsystem. There are alternatives to reporting energy used. One is to report actual energy input such as cubic feet of gas or kilowatt hours of electricity. The second is to report the specific quantities of fuels used to generate the produced energy forms in the subsystem.

The first approach has definite advantages. In this approach, mixed energy sources can be easily handled. For example, if the company wanted to evaluate the use of an electrically heated system compared to a natural gas system, they could evaluate the kilowatt hours to cubic feet of natural gas. Costs could easily be compared. The cubic feet of gas and kilowatt hours could be converted to megajoules for comparison. It is not uncommon to calculate the "precombustion energy" of a subsystem. The difference between combustion and precombustion of energy is what is included. Combustion energy is the energy contained in a unit volume, such as 175,000 BTU per gallon. If a gallon of this material was burned, it would release 175,000 BTU of energy. However, if obtaining the fuel took an additional 25,000 BTU of energy (drilling and production of the material, transportation, etc.), the total "precombustion" energy is 200,000 BTU. The inclusion of the precombustion energy is analogous to extending the boundaries of the system for energy. The inclusion or exclusion of the precombustion energy should be clearly stated. In most cases the energy used comes from commercial sources or is totally or partially generated on site. The inclusion of electrical energy from a public utility can be indicated as kilowatt hours and in dollars. The kilowatt hours can be converted to megajoules for comparison if necessary. Natural gas is usually obtained from a utility or gas supplier. The gas can be recorded as cubic feet and be converted to BTU or megajoules. Gasoline or diesel fuel for transportation is usually purchased from a retailer. Here too, the energy can be reported as gallons and converted, if necessary, to BTU or megajoules. In the case of most energy, it is considered coming from nonrenewable sources unless otherwise noted. The use of hydroelectric power would be an exception. Even in these cases, it is not uncommon for a utility to "mix" the sources of the electrical power from plants burning coal or oil with that from hydroelectric. An example of a power grid electrical source mix is

Fuel	%
Coal	5
Nuclear	20
Hydroelectric	10
Natural gas	10
Oil	4
Other[a]	<1

[a] Includes wood and waste to energy sources, geothermal, wind, but excludes cogeneration sources.

The theoretical conversion of kilowatt hours to common fuel units (joules) is 3.61 MJ/KWh. However, the user may use 11.3 MJ/KWh to reflect the actual use of fuel to deliver electricity to the user from the power grid.

Water volume requirements should be included in a life cycle inventory. How should water be included? The goal is to measure, per unit product, the amount of water that represents the water not available for beneficial use.

Water drawn from a stream, used in a process, and replaced in essentially the same quantity, location, and condition should not be counted in the analysis. This water is considered flow through. On the other hand, water taken from a groundwater source and discharged to the surface must be considered. The groundwater is not directly being replaced. Water from a utility and discharged to the sewer is considered. In the process of conducting the life cycle analysis, the analyst will consider the net amount of water consumed.

A life cycle analysis takes into consideration five types of outputs: airborne wastes, waterborne wastes, solid waste, products, and coproducts. Each of these categories of outputs is quantified during the study for each subsection and rolled up to higher levels as appropriate. In the past, some analysts have used just reportable wastes in the analysis. It is now the practice to report all waste discharges where the data can be obtained.

Air emissions are reported on a weight basis. The amounts should be the actual amounts of waste. It is permissible to report only the waste being emitted to the environment and exclude those being trapped or treated in a waste abatement system. However, the study will need to consider the treatment system as a subsystem and model it accordingly. If the waste treatment system is complex, it may be beneficial to treat it as a subsystem. Fugitive emissions and those from production or transportation are reportable. Typical atmospheric emissions include, but are not limited to

- Volatile organic compounds (VOCs)
- Sulfur oxides (SO_x)
- Nitrogen oxides (NO_x)
- Carbon monoxide (CO)
- Ammonia (NH_3)
- Lead (Pb)
- Chrome (Cr)
- Particulates

Carbon dioxide and water vapor are not usually included. If obtaining the data on these two substances is not too difficult, they should be included.

Waterborne wastes are reported in units of weight and include all substances. When possible, it should include all waterborne waste and not just those required to be reported by law. In general the broader the definition of wastes used and reported in the analysis the better the evaluation of the MANPRINT the product and process has on the environment. Any accidental discharges to the environment are reportable in the study. Some of the more commonly reported waterborne wastes and waste indicators are

- Biological oxygen demand (BOD)
- Chemical oxygen demand (COD)
- Suspended solids (SS)
- Settleable solids (SS)

- Dissolved solids (DS)
- Fats, oil, and grease (FOG)
- Phenols
- Metal ions (copper, iron, chrome, etc.)
- Anions (such as sulfides, chlorides, fluorides, etc.)
- Phosphates
- In some cases bacteria or virus
- Sludge or filtered solids

The above list is not all inclusive. Actual wastes will vary for each system depending on type of industry, process, and the range of water carrying wastes related to the product or process. For example, the waste from a food processing plant making corn chips is primarily corn meal. While not truly hazardous, much corn can be discharged. This will drastically raise the BOD, COD, and suspended solids levels. The owners of the plant might consider a treatment system enabling them to sell the waste corn meal as animal feed and reusing the water for cooling towers before discharge. The owners would now have a secondary product and get more "bang for their buck" with the additional use of the water.

Solid waste includes all solid material disposed of from all sources within the system being studied. Solid wastes are reported by weight. Depending on the comparisons, solid wastes may be converted to volume for comparisons in landfill disposal (see Figure 2-2). If volume numbers are used, weight should be reported as well.

It is sometimes desirable to segregate the wastes by type. This aids in determining which process may be capable of modification to reduce the wastes.

Solid wastes are also differentiated by production and postconsumer wastes. Postconsumer wastes are those wastes generated from product/packaging. This includes the packaging materials that are discarded by the user and the disposal of the product after its intended use.

Process wastes are those wastes generated during the manufacturing cycles of the product(s). This includes trim wastes, sludges, or any solid used to make the product that is not shipped as part of the product. All solid waste is considered, both hazardous and nonhazardous.

The depth that the analyst takes with wastes including solid depends on the effective boundary conditions set for the study. One may or may not include wastes from the refining of the fuels used at the production site(s) or for transportation, mining tailings from ore production, and slag from refining.

Products identified for the templates are defined by the system and subsystem being studied. At a top system level, the end product shipped by the company is the product. At subsystem levels, the "product(s)" are the output of that subsystem. In the case of the example at Acme Plastics & Plating, the end product is the plastic plated "widget" that is shipped. At the injection molding step, the product is the plastic molded widget. At this point, the widget

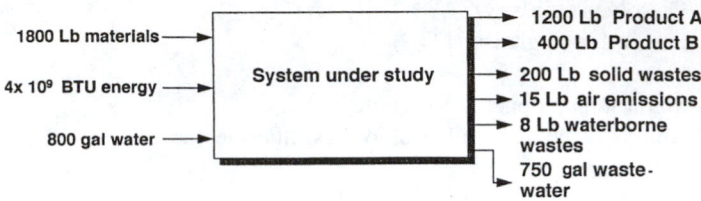

Figure 4-4.   Mixed product subsystem.

is not yet complete, but it is the output of that particular process; thus, it is the product for that subsystem. Each subsystem has a product and possible coproducts that are components or subassemblies of the final product. Anything else produced at each system or subsystem level that is not disposed of as a waste and gets further processing into another product or is sold as a raw material is considered a coproduct.

If a process or subsystem has coproducts, it is the custom to allocate all inputs and outputs to each product being produced. This allows for the evaluation of resources and wastes (Figures 4-4 through 4-6). While it is customary to consider all coproducts, each system being studied needs to be handled on a case-by-case basis. Coproducts are of interest only to the point where they no longer affect the primary product being considered. Subsequent steps past the point of interest to the primary product are beyond the scope and boundary of the study and need not be included past that point. If during the study sufficient details can be found about the coproducts to facilitate the study, then they may be summed at the point of interest. This reduces the time and effort required in modeling. For chemical processing, this can become quite complex. In cases such as these, mass allocation may be necessary. Where it is not possible to allocate separate coproducts, it will be necessary to lump them together as joint wastes. An example is the electrolytic treatment of brine. This produces sodium, hydrogen, and chlorine as coproducts. In this case, one might want to consider the emissions containing chlorine alone. However, the electrolytic cell also produces sodium and hydrogen at the same time. The cell cannot produce sodium and hydrogen alone. Chlorine will be coproduced. In this case all three may be lumped as a joint waste. Any allocation of power will need to be done on a mass weight allocation basis.

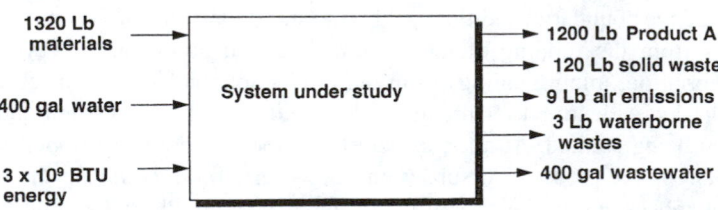

Figure 4-5.   Coproduct allocation — Product A.

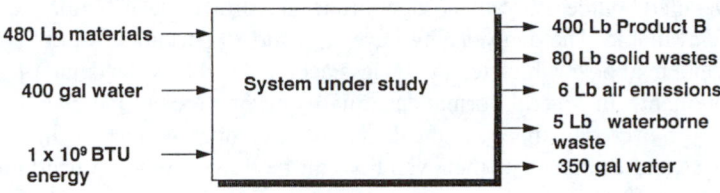

Figure 4-6.   Coproduct allocation — Product B.

Data quality greatly influences the end results. In the case of life cycle analysis, this is especially true. The development of uniform criteria is critical for the selection and reporting of data types and sources. Some basic objectives for data quality should be specified by the analyst based on the purpose of the study. In life cycle analysis, data can be considered being composed of two parts:

- Process and activity measurements that are amenable to standard statistical treatments
- Sets of assumptions and rules used for combining the data sets into a system

For purposes of this text, we will consider the process and activity. When collecting the data, careful consideration should be given to the age and frequency of the measurements. This is because technology or equipment can change and drastically affect the data. The frequency of the data acquisition is important to ensure that seasonal or other variability in the system is adequately captured. In addition, traditional indices of data quality such as precision, detection limits, completeness, and accuracy should be evaluated. The most accurate and recent data are desirable in a life cycle analysis. All data received must be carefully reviewed regarding the source and content before the data are used.

Much of the data being used will be from direct facility measurements or indirect estimates from published sources. Thus, accuracy is determined by the quality of the measurement and the estimation procedures used where all the variables were not known or controlled or by the averaging process used for representative systems. The use of industry data which have been averaged may be from different plants with different processes and equipment. They may also produce several products from the same process at each plant sight. Therefore, some engineering estimation and judgment is used when preparing the data for a particular product or process. Thus, the average data present for use may not be characteristic of any existing plant.

The quality of the data may be affected by confidentiality of the data. If the data are to be given to the public, it will tend to be more aggregated and more general in nature than if the data are generated to be confidential. Care must be exercised when using public or data from "industry" sources.

Detailed sources for all the operations in a life cycle inventory are not always available. The analyst may have to resort to periodicals, public databases, and textbooks which tend to be less accurate and lack the detail of direct measurements. In general, formal data quality criteria are not included in such sources. To overcome this issue the data may be compared to data from similar processes and estimates of their validity can be drawn. Where uncertainty is not understood, the analyst must note the condition, sources, and estimate of the degree of uncertainty.

Uncertainties in data are unavoidable. All data are not created equally. Some data have a greater impact on the outcome of the life cycle inventory than others. It is also not possible to establish the significance of any individual piece of datum to the final inventory set. The true test of whether an element of datum is significant to an inventory is the sensitivity of the final result to the elements inclusion or exclusion. The typical sensitivity analysis is conducted by evaluating the range of uncertainty in the input data and recalculating the model's output to see the result. Thus, the higher the degree uncertainty in strongly related variables, the less acceptable if the objectives of the study are to be met. As a rule, if an estimate of the true variability, such as measured statistical variance, is known, it should form a basis for the high and low range uncertainty estimates. The 95% confidence bounds are generally used for this purpose. For many inputs, the variability may not be known or may reflect variations of a process over time for which the analysis is being performed. It is usual in these cases to vary the input data by a range around its expected mean value. This may be as little as 3 to 5% for some variables and as much as an order of magnitude for others.

Order of magnitude estimates are sometimes used for setting the boundaries of an inventory. Deciding how far back to go, such order of magnitude estimates may be used to decide which inputs require a higher level of accuracy. In setting the boundaries, the analyst may want to determine if a process should be included or if order of magnitude estimates are acceptable. Running calculations with and without the data or a given process, it is possible to decide whether a variable has a large or small impact on the results. The analyst must keep in mind what the end use of the life cycle analysis is and who is the end user. In some cases, highly detailed sensitivity analysis is not needed or is minimized. For example, in one case the reporting of VOC in air emissions is adequate for the intended use. VOC is a measurement of a host of volatile chemicals in an air stream. The total VOC may be adequate. In another case the constituents of the VOC measurement will be needed as will the level of contribution of each one. In the first case, the user does not mind variability in the individual components. In the second case, care must be exercised to quantify each contributing chemical's presence and contribution to the total VOC measurement.

SPECIAL CONSIDERATIONS

Certain other conditions relative to data need to be highlighted. The first is the time period used for collecting the data. The time period needs to be long enough to smooth out any deviations or variations in the normal operating of the facility. A second condition is the geographic specificity. Natural resources and environmental consequences occur at specific sites but there are broader implications. It is important to describe the scope of interest in an inventory (local, national, international). Most inventories are done domestically and relate only to the country where the plant and customers are located. However, some studies are international in nature due to the location of natural resources, multiplant locations, or international use of the product(s). If no specific geographical data exist for a particular country when conducting an inventory, it is customary to use data from the U.S., Canada, Western Europe, or Japan where more accurate data most likely exist.

Care must be exercised to consider the technologies being used. For a life cycle analysis for a specific company, this can be relatively easy. For the use of industry data, the mix of technology may be more involved. It is wisest to use as much direct data as possible; however, the use of market share data to appropriate the data may be necessary. Accidental discharges or fugitive emissions need to be addressed. Whenever possible, fugitive and routine emissions and accidental emissions should be included. If data on fugitive emissions are not available, and quantitative estimates cannot be obtained, they may be omitted as long as this condition is duly noted. For accidental discharges, more frequent low-level discharges should be included. Low-frequency, high-level discharges, such as a major oil spill, probably would not be included. Tools other than a life cycle analysis may be more appropriate. In most cases, inputs and wastes from personnel at the site are not included. This includes such things as air conditioner emissions, lunch trash, wastewater from sanitary facilities, etc. Waste disposal of plant waste must be considered. Waste disposal in the total length of the life cycle must be considered. This includes such items as process wastewater treatment or disposal, product packaging wastes, and ultimate product disposal. The actual disposal method needs to be examined. For example, a "sanitary" landfill has emissions that are, or can be, hazardous to the environment. These include

- Methane gas (usually burned on site)
- Contaminated water from seepage and decomposition
- Sulfur compounds including hydrogen sulfide gas

The emissions from a landfill and how they are treated at the site are important. However, trying to determine how much is related to the product under study may be difficult. If the breakdown products are known, or if there isn't any noticeable breakdown for years, then determining the contribution by the product is not important. However, the product will take up space and that is important, especially if the company is comparing its product with competing product.

Specific Life Cycle Stages

5   RAW MATERIALS AND ENERGY INVENTORY

The boundary conditions establish how far back in the materials chain the analysis will go. In most cases the analysis starts with the aquisistion of raw materials and ends with the first manufacturing stage that refines the raw material. There are two kinds of raw materials: primary and secondary. The primary form is raw material that has not been recycled/reused. Secondary raw materials are those that are recycled/reusable materials. Secondary materials are given serious consideration beacuse of the following:

- Reduction in use of primary raw materials
- Process involved with preparation for use
- Emissions and environmental impacts of the processes

After the materials have been refined they are inventory. However, in some life cycle inventories, the inventory is considered raw material because the boundary conditions were set accordingly.

There is also the condition where the raw material for a process is provided as a premanufactured good. For example at Acme Plastics & Plating Company, the proprietary plating solutions used are purchased from a supply house. The impact of the manufacturing of the proprietary plating solutions may or may not have a major impact on the analysis, depending on what is being considered. In either case, data for the raw materials would most likely be summarized at the raw material's input point of the appropriate subsystem. If the analyst wished to go all the way to the mining of the appropriate ores, making of the acids, and compounding the solutions, it would be possible, but time consuming and expensive. Industry data would most likely suffice. The analyst may want to stop with the proprietary chemicals as the materials input to a product study and stop there. This would most likely be the case for a site study or one that uses chemicals such as plating. The major impact Acme would have on the chemicals it buys would be to change the plating processes if possible. This could reduce Acme's MANPRINT but may or may not do anything for the supplier.

Primary raw materials can be produced by cultivation, harvesting, and replenishment such as farm products or some types of wood or can be mined such as fossil fuels, ores, water, and air. All forms of transportation associated with the materials are also considered.

The inputs for the basic raw materials used in a life cycle analysis can be categorized into three groups.

The first is energy utilization. Detailed information on the type and mix of the energy sources utilized must be included in the database for each stage. Electrical energy should be reported as kilowatt hours. Other energy sources should be reported in appropriate units of use, such as gallons or liters of fuel or cubic feet of gas (see Appendix C). Energy use is then converted to megajoules and reported as such. The type of energy should be reported as renewable or nonrenewable. A review of primary energy sources is shown in Figure 5-1.

Materials that are consumed during the maintaining of the raw material source should be included, such as pesticides, fertilizers, etc. (Figure 2-1).

Various types of infrastructures should be included, such as construction and equipment. These include roads, buildings, drilling rigs, and equipment used to explore, mine, extract, or harvest the materials.

The outputs generated during the raw materials acquisition system can be generalized into a number of categories: air emissions, waterborne emissions, solid waste, other environmental releases, habitat changes, and the raw material itself.

One of the reasons to seriously consider the raw material acquisition phase as a subsystem in the life cycle analysis is the potential environmental impacts the acquisition process has on the environment. This includes agriculture runoff, oil spills, leaching of mine tailings, etc. These types of considerations

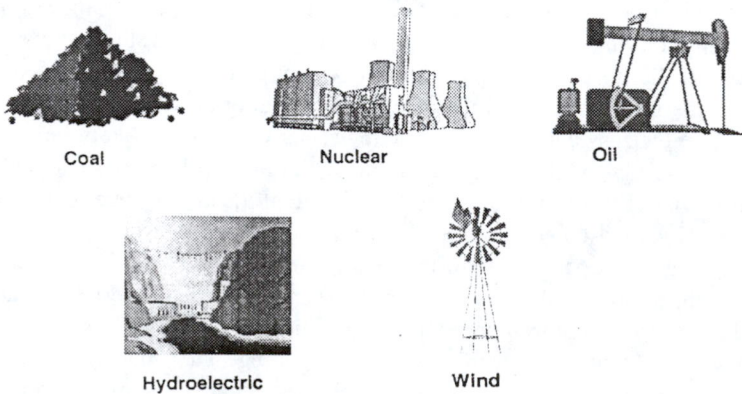

Coal   Nuclear   Oil

Hydroelectric   Wind

Figure 5-1. Primary energy sources.

can have a larger environmental impact than the sum of the rest of the product life cycle combined. Sometimes these impacts are difficult to quantify, such as aesthetic degradation and destruction of habitat. It is desirable to characterize these types of impacts in some way so they can be accounted for in the overall analysis.

Primary energy raw materials include petroleum, coal, natural gas, nuclear, wind and hydroelectric. A distinction is made between nonrenewable (coal, oil) and renewable (wind, biomass, hydroelectric) fuels. Fossil fuel raw materials inputs are reported as megajoules (MJ). This reflects their nonrenewable origin. A correction factor (precombustion energy) is applied to account for the energy required to obtain the raw material. Electrical energy is determined by stating the fuel inputs required to produce the net electrical energy consumed by the various processing stages. The distribution of electricity can be assumed to be off the national energy grid unless the analyst knows that the source is local. The consideration of the type of energy, the type of fuel used to generate it, and the amount used by the processes being studied is important. The cost of energy is of concern; however, the environmental impact of the energy source figures in to the overall MANPRINT of the processes and products being studied.

Secondary sources of energy include thermal energy derived from waste heat, combustion of biomass, combustible waste gases, combustible waste solvents and solids, and cogeneration from waste heat. The energy from these sources will be considered. To simply handle these sources, one can treat them as a process step (see Figure 3-6). Whether to call the energy from waste products a coproduct or an alternative energy source is up the analyst; however, it is important to remember to quantify it in the model.

Raw materials used in the study, excluding energy, can be in the form of solids, gases, and liquids. Water is usually categorized as an entity unto itself. While water is a renewable material, good quality water is becoming scarce. Wastewater may have to be treated before discharge, thus involving another process to be included in the model. The number of other raw materials that are extracted from the earth to produce the wide variety of products is surprisingly small. A typical list of basic raw materials would include those shown in Table 5-1. There is no simple way to characterize these materials. They are usually reported in units of weight. Gases can be in cubic feet or in weight units. The environmental sensitivity of the raw material must be identified and noted. Some materials are in short supply and are nonrenewable. The environmental impact of obtaining these resources should also be identified. Some involve processes and materials that are extremely hazardous to the environment. Since the use of renewable resources is deemed desirable, these materials should be flagged for special attention. Both renewable and nonrenewable resources must be accounted for in the inventory.

While a fundamental concept is to include all raw materials all the way to their source, it may not be practical. The importance of establishing the

Table 5-1 Typical Raw Materials List

• Barites	• Iron ore	• Water
• Bauxite	• Iron chromate	• Wood
• Biomass (cotton, wool, corn, soybeans, etc.)	• Lead	• Zinc
• Brine	• Limestone	• Gold
• Chalk	• Magnesium	• Silver
• CaSO$_4$	• Managnese	• Platinum
• Clay	• Coal	• Sand
• Copper	• Sodium nitrate	• Selenium
• Fieldspar	• Nickel oxide	• Oil
• Ferromanganese	• Uranium	• Natural gas
		• Rutile

appropriate boundaries is obvious. The extent of the boundaries and the magnitude of the study must be tailored for the individual application. The analyst must understand the intended use of the analysis and include the levels of detail required to make the report meaningful while not driving the client into bankruptcy.

DATA GATHERING

Because the data gathering is complex and critical, three basic approaches have been utilized for this stage:

- Primary data collection, where the raw materials producer directly describes how they produce their product and provides as much of the necessary data as possible.
- Secondary data, where published data such as articles, studies, and surveys are used.
- Assumptions, where the analyst makes assumptions about the parameters of the products use.

No single source is usually sufficient to conduct the life cycle inventory. In most cases, the data come from a combination of the above sources. Each source has its own limitations, advantages, and disadvantages as discussed below.

Primary Data Collection

Primary data collection can be done by statistical sampling. Here, the primary data collection is designed to capture a representative sample of data producers can provide on the parameters of their products. For example:

- Other materials used in conjunction with the product's raw materials acquisition and manufacture.
- How long the process has been in operation. Details about its operation, such as continuous service or interrupted, operating conditions.
- Method of by-product disposal or use for by-products.

Frequently, the producer cannot directly supply information on inputs and outputs. The producer may be able to supply information on how the product was obtained from which inputs and outputs can be derived. Most of this information the producer may have. Some of the information such as energy inputs and air/water/solids emissions and wastes may have to be derived by alternative means. This type of data collection takes several forms:

- Mail survey. A questionnaire is mailed to selected recipients
- Personal interviews
- Telephone survey

Each form has its advantages and disadvantages. This type of data collection can be difficult and expensive. National surveys have been conducted in a statistically valid fashion with methodologies that are well defined and codified. Most of this type of data collection is done by independent survey research companies.

Primary data collected that are statistically significant have common phases:

- Experimental design and protocols — A description of the objectives of the research, population of interest, and information to determine the sample size, the source of the list to be used for sample selection, sampling frequency, survey time frame, means of handling nonresponse, and other information.
- Survey instrument — The questionnaire used to collect the data, including instructions to the interviewer.
- Data tabulation and extrapolation — The tabulation and presentation of the survey and means to show that the sample is representative. A description of the methods used to extrapolate to the studied population.

When the source of the data is statistical sampling, even if well designed, there are issues that include the following:

- Cost
- Representativeness of the sample
- Bias in the questionnaire

- Respondent bias
- Nonrespondent bias
- Statistical interpretation

There are other sources of primary data. These include a smaller sample size and may not be representative of the total population. However, for materials that are produced by a small segment of the population, are used on a regional basis, or are geared for a certain segment of industry, the following techniques may be employed.

- Focus groups — This is a group of 8 to 15 producers that are assembled in a room to answer questions about the processes. This is usually conducted by professional marketing or survey companies. For raw materials suppliers, this isn't as difficult as one would expect.
- Suppliers of similar materials.
- Trade associations.
- Manufacturers of equipment associated with the process.

Issues associated with small groups or limited primary data collection are similar to those of the statistical sampling. The advantages include lower cost.

Secondary Data Sources

Secondary data sources are any published or unpublished data articles, reports, or studies relating to the materials that are available in published form. In many cases, good secondary data can be more comprehensible than limited primary data collection. Examples of secondary data sources include

- Trade, professional, or industry reports
- Articles in trade or professional journals
- Industry market survey and studies
- Federal government reports and statistical data documents and studies
- Reports of government rule-making hearings

The data from secondary sources may have been created for another purpose; therefore, some of the desired data may not be directly present. Some government reports may include most of the data needed for this phase of the life cycle inventory.

Issues with secondary data include

- Cost — Usually less than primary data.

- Bias — Depending on the secondary source, the data could be biased in favor of the product or manufacturer or against it. Care must be taken to ascertain any bias in the data.
- Timeliness — Secondary data are usually compiled at a time prior to the current study. It is imperative that the data be representative of current practices.

ASSUMPTIONS

Assumptions are used when primary or secondary data are not available. Assumptions can be based on data that are not specific to the process or service being analyzed. In most cases, the assumptions are used because data are not fully available.

It should be noted that the obtaining of data and the sources for all the stages of the life cycle inventory is pretty much the same. The analyst will use primary or secondary data or assumptions. The treatment of the data should be similar to what is discussed above, no matter what stage the analysis is in.

MANUFACTURING, FORMULATION, AND PROCESSING

6

The manufacturing, formulation, and processing step (hereafter known as manufacturing) in a life cycle analysis converts the raw materials into final products. This includes on-site storage and handling. The manufacturing step is considered complete when the product is transferred to distribution. All packaging required for the transfer of the product to the first distribution site is considered part of manufacturing.

Depending on the intended use of the analysis, the boundaries will include the raw materials to the processes. The end of the manufacturing step is at the distribution point. Figure 6-1 provides a conceptual view. Figure 6-2A indicates the "normal" type of multipath product flow. Figure 6-2B shows a just-in-time-type manufacturing flow. This is to indicate that both types of manufacturing systems have multistage operations and the boundaries must be made to accommodate them both. A manufacturer will be mainly concerned with the following:

- Principle raw materials
- Manufacturing of the product
- Energy
- Packaging
- Disposal/reuse/recycle
- Customer image

We will examine each of these as it relates to the design and manufacturing operations.

The boundary conditions will have defined the extent that the analysis goes back into the acquisition of raw materials. Certain materials will be of greater importance than others. These may include materials that are in short supply, environmentally sensitive, expensive, require large amounts of energy

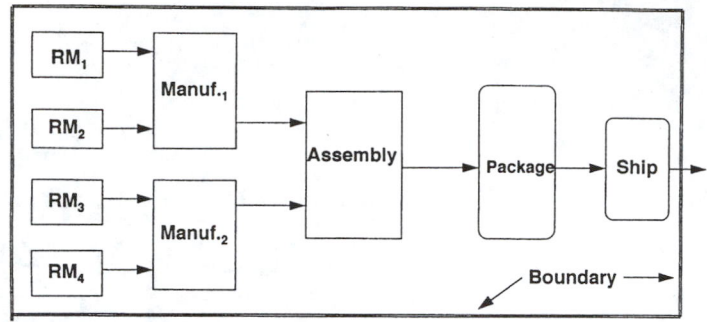

RM = Raw materials
Manuf$_x$ = manufacture of subassemblies 1 through X

Figure 6-1.   **Manufacturing boundary concept.**

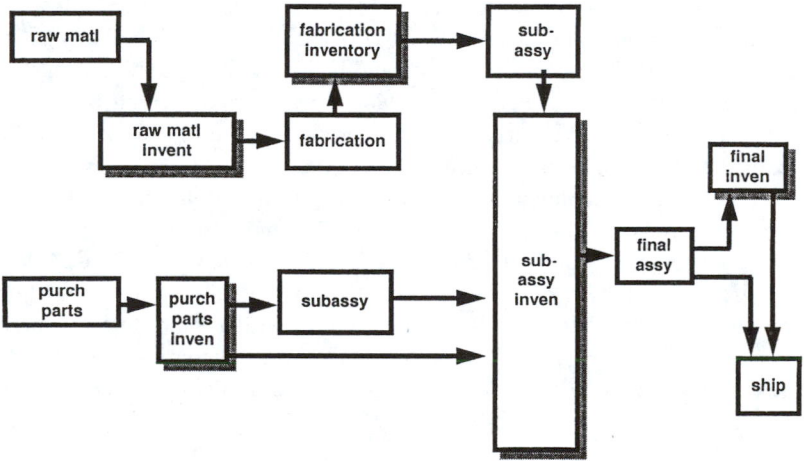

Figure 6-2A.   **Normal material flow.**

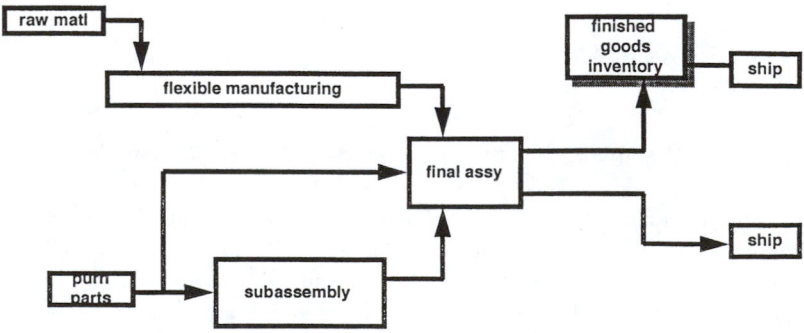

Figure 6-2B.   **Just-in-time-type material flow.**

to produce, and are major components of the products. Other raw materials may include recycled materials from other sources or from the company's own "take back program". Some of the recycled materials may be from an in-house waste treatment facility or a regrinding center in the case of plastics. The recycle facility must be included in the analysis. These facilities are usually considered manufacturing processes which have their products shown as material inputs to a manufacturing process. The net result is less actual raw materials used. This includes water (see Figure 3-7). In this concept diagram, the raw materials are transformed into subassemblies. These are then combined into a product. The figure also represents the packaging. This can include packing materials including foams and Styrofoam "popcorn", boxes, cartons, pallets, and shrink wrap.

As illustrated in Figures 6-1, 2-4, 3-5, and 3-6, products are usually made by a series of submanufacturing centers and assembly processes building to a final product. The overall assembly of a product consists of, as a minimum, the following:

- Raw materials for the product
- Energy
- Labor
- Materials, energy, and labor for tooling, fixtures, and assembly aids
- Manufacturing scrap and replacement materials and labor
- Rework materials
- Ancillary materials
- Packaging materials

The first three items on the above list have been discussed. It is paramount that the analyst consider the use of tooling. Tooling and fixtures used in manufacturing are of equal importance to an environmental life cycle analysis as the products. The use of exotic or environmental-sensitive materials for tooling is not uncommon. Management may not be aware of how extensive the tooling requirements are in a manufacturing environment. In a large percentage of cases, the need for special tooling can be reduced by modifications to product hardware designs.

Tooling can be made by machining metals such as alloys of steel, stainless steel, aluminum, brass, copper alloys, titanium, molybdenum, and others. Molds are made from the metals described as well as water-soluble salts, plastics, and plaster. Tools, molds, and fixtures are sometimes plated or coated to impart the desired properties to the tool. The use of chrome plate to increase wear resistance and coating with plastisol-type materials are examples. The use of the basic materials and subsequent coatings can have as great or greater impact on the environment than the actual product. Figure 3-6 illustrates the inclusion of tooling in the process flow modeling. The block in Figure 3-6 relating to tooling would be further expanded (blown up) to detail the processes involved. The modeling of the tooling process will divulge the quantities, types, and materials/coatings used and where used. This will allow for the appropriate level of analysis.

Scrap and associated replenishment materials must be considered. The level of scrap produced for a process adds not just to the environment, but to the overall cost of the final products. Facts that should be gathered about scrap include but are not limited to the following:

- Location of the generation of the scrap
- Quantity of scrap materials
- Type of scrap materials
- Disposition of the scrap (recycle in-house, third party recycle, land-fill, etc.)

Scrap is a subject of interest to most manufacturing companies. However, actual hard data quantifying scrap may be difficult to come by. There are two types of scrap. The first type is scrap from a process that involves the scrapping of a component that is a part of the product. The second type of scrap is surplus material left over from a process such as flashing from a molding process or metal turnings from a machining process.

The first type, product scrap, is the most costly. Product scrap may be handled as a contingency factor as a percentage of raw material purchased. This can mean the company overbuys raw materials and waste disposal or is constantly rushing to replace scrap materials. It also costs the company in labor. Most companies do not really know what their scrap rate is or where it is generated. The analysis of scrap will be of interest to senior management but may be difficult to develop. Scrap has a significant impact on the environment. It is a total waste, even when recycled. Someone put labor and other associated costs such as a percentage of the tooling costs into it before it was scraped. While recycling saves the material from the landfill and reduces the requirement for more raw material for the new user, the company generating the scrap has costs involved and must replace it with new material.

The second type of scrap is the process scrap. This includes materials used in the product that are scrap because of the nature of the process. Flashings and runners from molding operations and meat turning are examples. Spent solvents, rags, etc., are considered process wastes for the purposes of the analysis and not scrap. Process scrap should be noted and quantified. This can be of use to manufacturing and design engineers as well as management. While a certain level of process scrap may be mandatory by the nature of the process, efforts may be made to minimize it through design, tooling, or process changes.

When a company makes a design change for a performance requirement, it must look at the total ramifications. For example, automobile manufacturers were required to improve fuel efficiency. One aspect of the reduction was to use plastic parts, where possible, as a substitute for steel. This change helped reduce the average automobile curb weight 25%, from about 4100 to 3100 pounds. This raised the fuel efficiency from about 14 mpg in 1973 to about

28 mpg in 1995. The use of plastics and downsizing provided for the improvement in fuel usage. While the use of plastics helped reduce fuel usage, most plastic used in automobiles is not recovered. The use of plastic, therefore, increases the solid waste generated at disposal. The plastic generated at the shredder operation is called "fluff". The fluff has been increasing constantly. If the plastices cannot be recovered and recycled, in a practical manner, before or during the shredding operation, the shredding process may become uneconomical. Thus, to solve one problem, the automakers have created another one. In this example, the automakers have changed the type of scrap at their plant sites from primarily metals to a mix of metal and plastics as well as created an environmental impact at disposal.

The disposal practices for scrap should be included in the life cycle inventory. The method and quantity of scrap materials being disposed of has an affect on the MANPRINT of the processes and ultimately the associated product. If the materials are being added to a landfill, the volume as well as the costs are important. The disposal in a landfill has the greatest environmental impact of the three methods noted. If the material is considered hazardous, then additional considerations must be noted:

- The use of class I landfills
- Special transportation and packaging issues
- *Long-term liability*
- Limited on-site storage
- Special on-site storage requirements
- Additional training for employees

The long-term liability of hazardous waste disposal/treatment needs to be fully documented and evaluated in the life cycle inventory. There is not just the present impact to the environment but serious long-term considerations and potential litigation.

The on-site recycle of scrap materials will be included in the model as a process. This might be as simple as a regrinder for thermoplastics for reuse in molding. The recycled scrap material may end up in the original product, such as used ABS plastic at Acme Plastics & Plating being returned to the injection molding machine for reuse. It may be reused in a different product. A major automotive company makes closeout panels for one of their model cars from the material from minivan bumpers.

Off-site recycling can include the use of the scrap materials by another company. It will be indicated on our inventory as a scrap/waste that is used for raw material for a third party. It will have a lesser impact on the environment. Examples include an automotive firm where old tires are becoming foot pedals and old soft drink bottles are truck grills. Serious consideration should be noted in the life cycle inventory for the company's use of recycled materials as raw feed stocks. Companies that make copiers and printers usually have a "take back" program. The end user, in these cases, returns the used cartridge to the manufacturer via prepaid postage in the box the cartridge came in. The manufacturer then remanufactures a cartridge for resale by reloading and servicing the returned item. This reduces the overall impact of the company's MANPRINT by utilizing a "waste" from another company or consumer, removing it as a source of landfill material and reducing the MANPRINT of their product.

As noted above, most companies do not have a good handle on scrap. Even more surprising to the nonindustrial person is the fact that most companies do not have a handle on rework or repair either. Rework differs from repair by its definition.

- Rework returns an item that does not meet the drawing or engineering requirements to a condition that does meet the requirements.
- Repair means the item is made to comply with the drawings but does not meet the requirements. Using fill and paint as a means of repairing a dent on a car fender is an example. It looks like the original, does the job, but is not the same as the original.

Repair and rework at a manufacturing facility need to be included in the life cycle inventory. This is because the repair/rework processes require materials to be scraped and replaced and special tooling may be required as well as possible use of chemicals for cleaning, flux, solder, etc. The rework/repair cycle causes the generation of scrap, and these materials must be resupplied. There is a definite impact on the environment and associated costs involved.

Ancillary materials include cleaning agents, cutting oils, grease, rags, flux, mold release, and other similar products utilized to maintain the processes but do not become part of the product. Trace amounts below the level of concern should be noted as ancillary materials.

MANUFACTURING OUTPUTS

The various manufacturing steps ultimately result in the production of products, coproducts, and wastes. A product is a marketable commodity and the principal outcome or result of the manufacturing process. It is not uncommon for a manufacturing process to have one or more coproducts that could be used for alternative purposes. Examples include

- The making of corn chips. The principal product is corn chips; a coproduct is the waste corn meal that is sold as animal feed.
- The "boiling" of beef hides and bone for the making of gelatin. The first boil is the principal product; it is used for gel used on good-quality film. The second boiling produces the gel used on photographic paper to develop photos. The third boil is used for cooking

gelatin and food stuff. The last boil was used for glue manufacture. The resultant waste products are used as bone meal and fertilizer.

- Meat processing produces meat (beef), the principal product. It also produces beef tallow used in the production of soap, leather, and bone that is used for bone meal. The remaining materials are considered waste as they have no intrinsic value or alternative function and are disposed of in the environment.

Because of the multiple products that may be generated, it is important to clearly separate products early in the life cycle inventory so that the relative proportion of waste and energy may be properly assigned. For the corn chip producer, the energy and wastes would be apportioned over the corn chips and animal feed.

To assist in the quantification of relevant impacts of raw materials, energy, and waste emissions, it is common to assign weighted averages allocated on a weight basis. Exceptions to this should occur only when actual source data are clear enough to distinguish that an impact is caused by one product or a coproduct.

Wastes are to be considered at point of emission and after any waste treatment. That means wastes are evaluated only after processing through all in-line or on-site waste treatment and control systems to reflect direct emissions to the environment. Some wastes escape during the manufacturing cycle and never see the waste treatment abatement or control system. These are "fugitive emissions" and must be considered in the life cycle inventory. These types of emissions may be difficult to quantify. In these cases, the identification of the emission and a qualitative estimate may be required.

The manufacturing cycle can be simple or extremely complex. Each step in the manufacturing process must be scrutinized and the outputs noted. The wastes from all the subprocesses are totaled for the evaluation of the total wastes. However, it is imperative that the analyst review and report on the outputs, including wastes, at each stage of the manufacturing process. By reporting the wastes at each submanufacturing step, management can get a better look at the details of their company's MANPRINT and identify processes for further evaluation for MANPRINT reduction. Care must be taken to appoint all outputs to a proper input. The product output of a subprocess must be indicated as the input to a subsequent process. The wastes are shown as an input to a waste treatment/handling/storage or disposal process. The use of the modeling scheme depicted in Chapter 3 allows modeling down to the most appropriate level in an efficient manner.

There are practical limits to the amount of time, cost, data availability, and variations in technology throughout industry to necessitate the use of certain assumptions in the life cycle inventory. It is important to the quality of the analysis that these assumptions be made explicit. These assumptions should be understood and analyzed to ascertain the sensitivity of the assumption to the life cycle analysis.

Most people do not think about the processes involved with packaging and shipping as part of manufacturing. When the products are completed they must be prepared for transportation. The packaging of the product therefore involves three considerations as a minimum:

- Internal material handling
- Product packaging
- Storage and shipping

Internal material handling includes *special* racks, fixtures, and enclosures used during manufacturing to build and protect the hardware. It is important to note the quantity and type of packaging involved, materials used, an estimate (if not directly quantifiable) of energy consumed, and wastes generated (see process model Figure 3-6).

Product packaging must satisfy at least the following conditions:

- Product protection from handling
- Protect the product during storage
- Look pleasing for the end user
- Identify the product and manufacturer (and other legal requirements)

Product packaging may be almost any size, shape, and form, constructed out of almost any material(s). For example, the product may be held on a cardboard "platform" using a vacuum-molded clear plastic covering, shrink wrap of cardboard boxes containing the product, or be placed in a cloth-lined wooden box or anything in between. The use of foam for shock absorption or containment is especially noted for possible environmental considerations during the analysis. The important issue is to consider the packaging as a process and measure all the inputs and outputs in the same manner as other process.

Product shipping must, as a minimum, satisfy the following requirements:

- Protect the product during warehousing and internal transportation, i.e., forklifts, etc.
- Protect the product from the elements
- Protect the product during loading and shipping
- Protect the product until unpacked for sale or by the end user

The same considerations must be made for the final shipping of the product. The packaging for shipping is another manufacturing process and should be modeled as such.

Packaging is a science unto itself. The analyst must consider the processes involved in product and shipping/distribution packaging as unique processes that are an integral part of manufacturing.

SENSITIVITY

Energy variables should be noted. This includes electricity, which, if possible, should have its source noted. Other factors include whether it was from renewable or nonrenewable fuels. It will include whether the source was domestic or imported.

The variations in process waste releases and how they are accounted for is noted. If the waste is released to a treatment facility (on site or off site), it will be noted as well. The type of waste treatment facility, technology used, level of treatment, and location will be noted. The above is important as the type of technology used for a specific type of waste as well as its effectiveness and may vary within an industry. Whenever possible, the treatment plant should be modeled just like a manufacturing process. This allows all the appropriate materials, energy, and outputs to be measured or accounted for.

A number of authors in the life cycle literature refer to energy and materials balances. These are, simply put, attempts to quantify the amount of energy and materials used in a given process. The idea is basically to quantify all the inputs to a process and then quantify all the outputs for each material and amount of energy used. To a certain extent this can be done. However, with complex products and the relative size of most companies, the quantification of this is almost impossible. The most practical approach is to use the modeling system from Chapter 3 and model the manufacturing operations to the lowest level practical. Then, with the best data available, plug in the numbers for all the inputs and outputs. There will be some pieces of data that are not available or measured. The idea is to account for everything going into a process and the outputs as much as possible. If fugitive emissions are taken into account, this may be the best you can do, especially if cost is important. Parts of the product may be subcontracted, such as fabricating printed wire boards (PWBs) used in electronics. It will be necessary to model the PWB fabrication process and allocate energy and materials used for your particular PWBs. Assumptions may be necessary. Care must be used in this example because technology and equipment may vary from one supplier to another. The use of material and energy balances were commonly used in the remediation of hazardous waste sites. It became evident to more enlightened regulators and owners that materials and energy balances were the brainchild of consultants. The real value of an extensive energy and material balance is dubious at best. There are usually more assumptions than facts in them. The cost of them must be weighed against the expected result. Conducting the process modeling and plugging in the appropriate data will provide a balance, that is, in most cases, adequate to conduct a life cycle analysis and be relatively accurate. The use of the models will also highlight what data are not available and the assumptions that are made.

Care must be taken not to compare the processes of different-sized facilities. The use of different processes, technologies, and equipment will also slant comparisons. An understanding of this concept is important, especially when discussing the results with management. Another area to be sensitive about is the capacity of a plant. Executives sometimes misapply the term "capacity". They mistakenly refer to capacity as output. Capacity means the amount of output the company could get running a specified number of shifts and specified number of days with the existing equipment. Output means the output under existing condition. Utilization is the ratio of the two. For example with the present suite of equipment, Acme Plastics & Plating would have one number for a capacity at 1/8/5. This means one shift at 8 hours per shift for 5 days a week. It would be different at 2/8/5, (two shifts at 8 hours each for 5 days per week). 2/8/6 means two shifts of 8 hours each, 6 days per week. It is important, when discussing capacity, to take into consideration such things as experience, age of equipment, yields, and accident rates.

The usefulness of the analysis gains importance when the magnitude of the variables is tested for their effect on the results and the likelihood of variation is considered. The probability of change as well as the magnitude should be reflected in any sensitivity analysis performed.

The use of recycled materials must adhere to a constant convention if the material is brought into the manufacturing system, especially if different recycled materials are used at different steps. The inventory must include energies, other materials used in recycling, and waste releases connected to the recycling and distribution processes. For closed-loop recycling, the processes related to recycling must be considered as a step in manufacturing.

DATA GATHERING

Because data gathering is complex and critical, three basic approaches have been utilized for this stage:

- Primary data collection, where the manufacturer directly describes how to produce the product and provides as much of the necessary data as possible.
- Secondary data, where published data such as article, studies, and surveys are used.
- Assumptions, where the analyst makes assumptions about the parameters of the product's manufacture.

No single source is usually sufficient to conduct the life cycle inventory. In most cases, the data come from a combination of the above sources. Each source has its own limitations, advantages, and disadvantages as discussed below.

Primary Data Collection

Primary data collection can be done by statistical sampling. Here, the primary data collection is designed to capture a representative sample of data manufacturers can provide on the parameters of the products. For example,

- Other materials used in conjunction with the product's and coproduct's manufacure.
- Frequency of product repair or maintenance (warranty).
- Length of time the processes used for manufacture have been in production, details about their operating conditions, etc.
- Other coproducts produced from the same set of processes.
- Method of product disposal when the end user was through with the product.

Frequently, the manufacturer cannot directly supply information on inputs and outputs. The manufacturer may be able to supply information on how the product was produced and from which inputs and outputs. Most of the details of this information the end user may not be aware of unless the manufacturer has an outstanding accounting system. Information from the industry can sometimes be used, but with caution. Not all producers of a product use the exactly the same processes. This type of data collection takes several forms:

- Mail survey. A questionnaire is mailed to selected recipients.
- Personal interview.
- Telephone survey.

Each form has its advantages and disadvantages. This type of data collection can be difficult and expensive. National surveys have been conducted in a statistically valid fashion with methodologies that are well defined and codified. Most of this type of data collection is done by independent survey research companies.

Primary data collected that is statistically significant have common phases:

- Experimental design and protocols — A description of the objectives of the research, population of interest, and information to determine the sample size, the source of the list to be used for sample selection, sampling frequency, survey time frame, means of handling nonresponse, and other information.

- Survey instrument — The questionnaire used to collect the data, including instructions to the interviewer.
- Data tabulation and extrapolation — The tabulation and presentation of the survey and means to show that the sample is representative. A description of the methods used to extrapolate from the studied population.

When the source of the data is statistical sampling, even if well designed, there are issues that include the following:
- Cost
- Representativeness of the sample
- Bias in the questionnaire
- Respondent bias
- Nonrespondent bias
- Statistical interpretation

There are other sources of primary data. These include a smaller sample size and may not be representative of the total population. However, for products that are made by a small segment of industry and are used on a regional basis, the following techniques may be employed.

- Focus groups — This is a group of 8 to 15 manufacturers that are assembled in a room to answer questions about the product. This is usually conducted by professional marketing or survey companies.
- Suppliers of raw materials — For example, soap manufacturers for washing machines.
- Trade associations.
- Manufacturers of equipment associated with the product.

Issues associated with small groups or limited primary data collection are similar to those of the statistical sampling. The advantages include lower cost.

Secondary Data Sources

Secondary data sources are any published or unpublished data articles, reports, or studies relating to the product that are available in published form. In many cases, good secondary data can be more comprehensible than limited primary data collection. Examples of secondary data sources include

- Trade, professional, or industry reports
- Articles in trade or professional journals
- Industry market survey and studies

- Federal government reports and statistical data documents and studies
- Reports of government rule-making hearings

The data from secondary sources may have been created for another purpose; therefore, some of the desired data may not be directly present. However, it may provide valuable information on the processes used to make the product and methodologies used to estimate the current data. Some government reports may include most of the data needed for this phase of the life cycle inventory.

Issues with secondary data include

- Cost — Usually less than primary data.
- Bias — Depending on the secondary source, the data could be biased in favor of the product or manufacturer or against it. Care must be taken to ascertain any bias in the data.
- Timeliness — Secondary data are usually compiled at a time prior to the current study. It is imperative that the data be representative of current practices.

ASSUMPTIONS

Assumptions are used when primary or secondary data are not available. Assumptions can be based on data that are not specific to the product or service being analyzed. For example, data on taking photos with a particular camera could be extrapolated to a single photo, or washing a load of dishes in a dishwasher could be extrapolated to washing glasses. In most cases, the assumptions are used because data are not fully available.

7　DISTRIBUTION AND TRANSPORTATION

Virtually every product manufactured has a distribution and transportation stage in its life cycle. Any life cycle analysis that does not include this stage is deficient.

The most common attribute for distribution and transportation is that it involves a change in location or physical configuration of the product, not a transformation of the product. Boundaries are not always sharp in this area and therefore must be defined at the onset of the analysis. An example is the freezing of meat or food products during transportation. If it is already frozen at the plant, moving frozen meat or food products is just transportation under special conditions. If it is frozen during transportation then it is in this fuzzy area as to definitions. The making of concrete in the truck on the way to a job site is another example.

For the purposes herein, we have defined distribution and transportation as follows:

- Distribution is all nontransportation activities carried out to facilitate the transfer of manufactured products from their final manufacture to their ultimate end user. This includes activities in a warehouse or retail establishment. This includes warehousing, wholesaling, retailing, and support activities carried out at these locations such as repackaging.
- Transportation is the movement of energy or materials between operations at different locations. (This includes moving raw materials from their source to the production facility, moving finished goods to a warehouse, moving products from the warehouse to retailer and ultimately from the point of sale to the end user, and the movement of electricity from generator to user.) This does not include the transportation or movement of product within the production facility during manufacturing. That movement is a process in manufacturing.

Table 7-1 Common Transportation Modes

Pipeline	Electrical power lines
Airplane	Railroad
Truck	Automotive
Barge	Freighter
Tanker	Electronic

The quantification of inputs and outputs related to transportation must be clearly mapped out. When fuel for the transporting equipment is considered, the environmental impact of the production and delivery of the fuel along with the associated costs must be taken into account. The relative amount of emissions and their contribution to the MANPRINT of the product must be included.

In the analysis of both transportation and distribution, where appropriate, environmental controls such as maintaining temperature and humidity are to be included.

For each mode of transportation, measures will be described that with a given quantity of product will allow the determination of energy, materials, and waste requirements. Examples of common transportation modes are given in Table 7-1.

The purpose of a life cycle inventory is to assess all the requisite process to get a product from raw material to the intended end user and assess the environmental impact. We can consider the transportation step a "process". Like the processes used and modeled in manufacturing, a simple model can be fabricated for transportation. A top-level model is shown in Figure 7-1. The model has inputs to be determined and quantified and outputs that need to be identified and quantified. These data can then be used to ascertain the environmental impacts and what possible improvements could be made. The shipping or transportation cycle might involve multiple operations. The steps should be modeled as with production (Chapter 3). The sublevel that the model goes down to will depend on the boundary condition set and the amount of detail required. An example is shown in Figure 7-2. A similar set of models could be made for pipeline or electrical transportation systems (Figures 7-3 through 7-5). A fundamental difference in transportation systems is that the movable transportation system (trucks, boats, airplanes) moves batch lots and is thus called a batch carrier. The fixed transportation system (pipelines, electric cables, ducts, etc.) moves continuous amounts of products. Batch carriers may have integral containment (tank truck) or separable containment such as container truck or railroad box car.

The previous chapters have discussed the requirements for the early establishment of boundary conditions. The area of transportation and distribution is no different. Hereto, they must be explicitly defined. There is no "right or wrong" set of boundaries. The boundaries are determined by the intended use

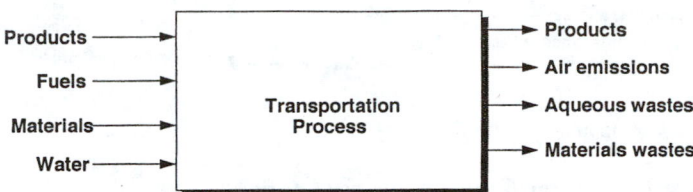

Materials = oils, lubricants, antifreeze, rags, ropes, tires, filters, belts, hoses, etc.

Figure 7-1.   Transportation model format.

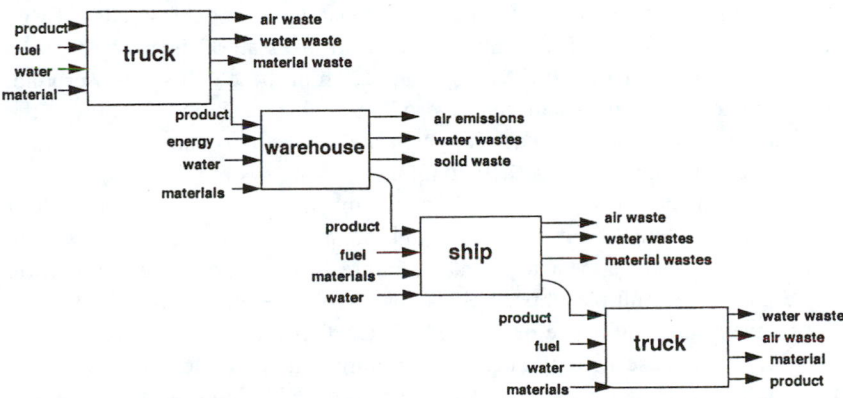

Figure 7-2.   Multilevel transportation.

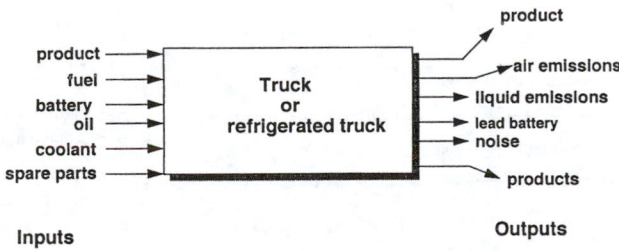

Figure 7-3.   Examples of truck transport.

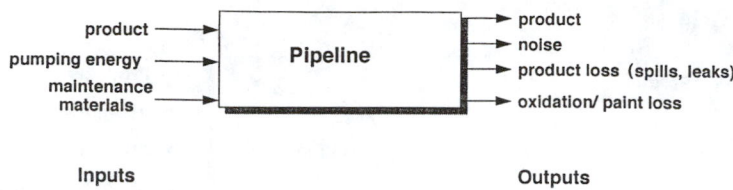

Inputs   Outputs

Figure 7-4.   Typical pipeline transportation.

of the study and the relative costs associated with expanding boundaries. For example, the use of trucks in our model might only incorporate the operation of the truck. At a higher level, it could be expanded to include maintenance and spare parts and at a still higher level it could include the making of the truck. The usual levels include the basic to first levels only. In some cases, we need to consider the inclusion of multiple products during a life cycle analysis. Such things as home computers come as separate components (CRT, CPU, keyboard, cables and printers, modems, etc.). It takes all of the subsystems to make a computer that the user can operate. Each of these items was a product at the factory. When considering a product such as a home computer, the analysis must have all the requisite parts, and it needs to consider the transportation and distribution of each component in the system.

The distribution step in a life cycle inventory sometimes gets short changed. After all the effort in the manufacturing stage and transportation, distribution doesn't appear to spark the interest of analysts. The distribution stage can have multiple opportunities for environmental improvements and does have a large influence on the MANPRINT of the product.

The warehouse activity accounts for numerous areas for environmental impact. These include recycle or disposal of the packaging used for the initial shipping. The methods of storage and the use of motorized equipment need to be assessed. Repackaging for distribution to retail outlets, pallets, and water and energy requirements are necessary inputs. Outputs will include heat, noise, office, air, and waster wastes as well as the product.

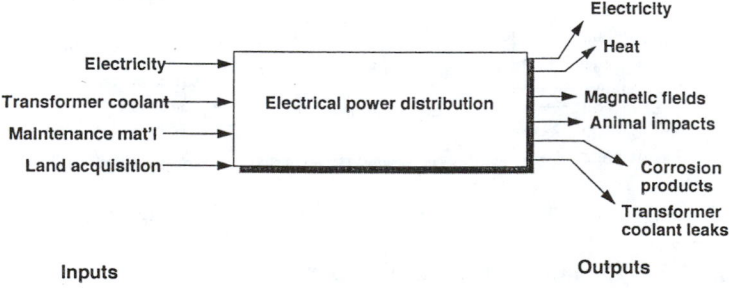

Inputs   Outputs

Figure 7-5.   Electrical power transportation and distribution.

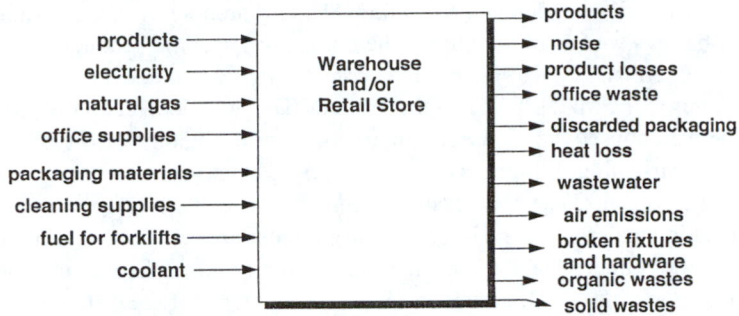

Figure 7-6. Warehouse and retail model.

The retail outlet has most of the opportunities as the warehouse plus the use of boxes, bags, and other consumer packaging for consumer transportation of the product. Figure 7-6 is a top view of the distribution model. It can be broken down to the level necessary to meet the boundary conditions of the inventory being undertaken.

The types of environmental impacts the distribution system has are drastically different from those of manufacturing. At the retail site, for example, there is the normal handling of the products by the staff as well as by customers. This will contribute to the breakage and product loss. The parking lot itself has an environmental impact that is not shown in Figure 7-6. The inclusion of the parking lot will depend on the boundary conditions and items such as who owns the parking lot, whether it is shared by multiple retailers, etc. The parking lot is considered a point source for pollution. Parking lots can contain, as a minimum, the following:

- Oil on it from leaking vehicles
- Antifreeze
- Solid waste from careless or inconsiderate customers
- Animal waste
- Grease
- Broken carts, fixtures
- Discarded automotive parts
- Dead animals
- Possible industrial emissions (directly or from precipitation from the air)

When it rains or snows, these materials can run off into creeks, streams, and rivers or slowly soak into the ground and groundwater. Because of this the U.S. EPA has regulations governing these areas. However, the analyst must consider whether to consider the parking lots in the study or omit them. If they are omitted, there should be a clear explanation as to why they were excluded and any overt observations.

The inclusion of detailed modeling of the warehouse/retail distribution should be reviewed as to the impact the company's specific product is having on the MANPRINT at this step and, conversely, how much the retail step contributes to the MANPRINT of the product. This will depend on the product and the warehouse/wholesale/retail network. It will obviously different for a refinery distributing gasoline as compared to an electronics company selling personal computers. The packaged electronic hardware will have the container or packaging materials as the largest environmental impact during distribution, whereas the refinery has the gasoline with virtually no packaging except the pipelines, tanker trucks, and underground storage tanks. Leaks at any part of the transportation and distribution phase could be catastrophic. The boundary conditions will define the extent that the distribution step is modeled. When looking at the distribution step, the inputs and outputs must be attributed for each product. This insures that each product gets its share of the inputs and outputs, such as wastes associated with that particular product. In the case of the gasoline, the inputs and outputs at the gas station are appropriated over the amount of gasoline, oil, grease, spare parts, and anything else sold. Since gasoline and oil are usually the greatest volume of materials sold, most of the inputs and outputs are attributed and distributed to them. The sale of maps, for example, is small by comparison. This can get more complicated for a personal computer sold at a Target store, for example. Target sells thousands of products; the computer is only one and is not sold in the same volume as towels, clothing, or hardware and automotive products. Trying to attribute the inputs and outputs at a Target store to the personal computer might be extremely costly and almost impossible to accomplish. To add to the problem, Target has different-size stores and sells different products in different parts of the country. There isn't much call for snow shovels in Southern California, but there is in Minnesota. In this example, it might be better to estimate the impact or consider it negligible. An example of a simple way to handle the computer sales of Brand X computers at Target would be to approximate the cost and amount of utilities, etc., for all products sold in a year. Then calculate the per cent of total volume or per cent sales dollars the Brand X computers are of the total sales or amount of merchandise sold. The analyst would then allocate the appropriate amount of utility usage to the Brand X computers. This method may not be pretty, but it saves much time and money. The amount of energy in kilowatt hours for electricity, and terms or BTU of gas, can be found in utility bills or from the local utility companies. These units can be easily transfered to megajoules.

The data requirements for the distribution and transportation stages are driven by the specific inputs and outputs associated with each transportation and distribution operation that is defined in the life cycle model. It is imperative to use the model technique. The data collected will vary depending on the product and transportation-distribution system used.

The life cycle inventory describes the total quantities of the inputs and outputs for each transportation and distribution operation. The quantities of each of the inputs and outputs are calculated from knowledge of the amount of product that is delivered to the end user. These quantities are usually reported as items per unit level of each of the transportation and distribution operations. An example includes the energy required to move a kilogram of product over 1 kilometer. In this case the analyst would convert the fuel consumption from gallons to energy. Therefore, the reported quantity would be in megajoules per kilogram-kilometer.

Some further examples would include but are not limited to

- Megajoules per kilogram-kilometer.
- Engine pollution emissions NO_x/Kg-Km (NO_x is reported in kilograms).
- Discarded and additional packaging materials (weight of materials in kilograms per kilogram of transported-distributed product).
- Product loss rate (kilograms of product loss per kilograms of product moved or distributed).
- Gallons or kilograms of water used per kilogram of product.
- Supplies required (kilograms of wood pallets per kilograms of product moved or sold).

Obtaining the data for the transportation and distribution step can be somewhat difficult. Fortunately, some of the data exist as factors and are available in the public sector. Trucking/shipping companies have some of it; regulatory agencies can provide some as well. The U.S. Departments of Transportation and Energy have data available to the public. Data relating to energy use and environmental emissions can also be obtained from regulatory agencies in most areas of the country. Some of the data are proprietary and may or may not be released to the analyst. Portions of the data will need to be developed. The models used for transportation can be somewhat generalized as the basics are the same for a wide range of products. Simple models can be constructed for the different types of transportation, see (Table 7-1). For the model of a truck moving between two locations, the model will need, as a minimum, the amount of energy used to transport the product, the amount and type of product, emissions released, noise and heat generated, and the routing schemes used. The model can be further refined as required. The extent to which this is done will depend on the boundary conditions which were established based on intended use of the study and cost considerations.

8   USE — REUSE — MAINTENANCE

How the product is used, maintained, and reused will have an important impact on the environment. Much of the impact is predetermined by the engineering design of the product. The materials of construction, product packaging, and required maintenance are predetermined by the design of the product and the design engineers. Understanding of the customer requirements and how the product will actually be used by the end user is critical for the designer. If the designer is to reduce the environmental impact of the product, he or she must understand the total life cycle of the product. This chapter deals with specific issues related to the end user stage of the product's life cycle. To accomplish this, we need to address, as a minimum, the following:

- Boundaries
- Activities in the user-reuse-maintenance (URM) stage
- Presentation of inputs and outputs
- Modeling
- Data sources
- Recommendations

The URM stage of the life cycle takes place after the distribution step and ends at the point where the end user discards the product or it enters a waste management system. This stage can be as complex as the manufacturing stage. The boundaries for this stage are illustrated in Figure 8-1. The figure shows that the URM stage begins when the end user takes possession of the finished product, including the packaging, and uses or consumes the product and ends when the used product is discarded. This definition can encompass all products from commercial products, defense weapons systems, and equipment. Everything from paper clips to aircraft carriers fits into this system.

The actual delivery of the finished product to the end user is considered part of the distribution system. This includes any means of delivery including the U.S. mail, UPS, Federal Express, private shipping service, boat, pipeline

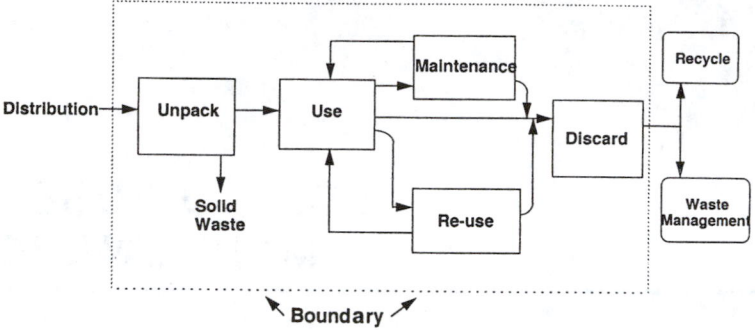

Figure 8-1.   User-reuse-maintenance stage boundaries.

or airplane (remote locations), etc. The product delivery can be from a take-out restaurant, retail outlet, mail order company, or other distribution source.

The first step in the boundary is the unpack step (Figure 8-1). During this step, the finished product is removed from its packaging and prepared for use. Wastes from this step include the packaging and shipping materials (i.e., shipping boxes) used to protect the product during transportation, storage, and distribution. The amount of discarded materials will be noted per actual unit of finished product. This is usually provided in units of weight.

The second step is the actual use of the product. In this step the product is placed in service or is consumed by the end user (Figure 8-2). This step includes all activities involved in the use or consumption of the product. Examples include the use of a refrigerator; power for a hair dryer; natural gas for heating water and the water for washing with soap, natural gas, and electricity for a furnace; water, power, and chemicals for a swimming pool; water and power/gas for the preparation of food; gasoline, water, antifreeze, grease, and oil for an automobile, etc. Each material, energy, and water input and output requirement will be quantified for inclusion in the model. Obviously, these will vary drastically by the nature of the product. The materials, energy, water, gases, or air consumed during use of the product must be identified and quantified in units that are compatible for the purposes of the study.

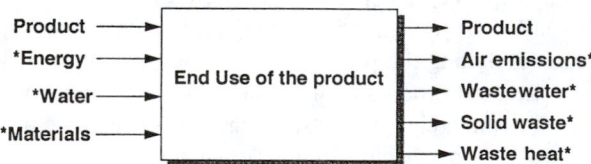

* = The extent to which these items are considered depends
on their existence based on the nature of the end product.

Figure 8-2.   Use of the finished product model.

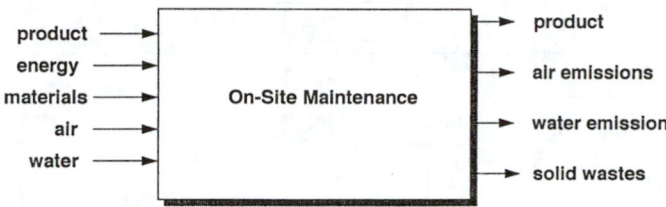

Includes cleaning, supplies, misc. transportation effects.

Figure 8-3. On-site maintenance.

Maintenance is considered for most products. This can include grease, oil, filters, tires, paint, etc., for automobiles and trucks, polish of silverware, replacement cartridges for copy machines and printers, shingles for a house roof, cleaning units for VCRs, and parts and lubricant for industrial machines. The maintenance step may occur at the site of the end user or at an off-site location. The maintenance may be accomplished by the end user or people on the user's staff (Figure 8-3) or by a contract repair facility. Off-site maintenance (Figure 8-4), therefore, includes transportation to and from the maintenance site. All materials, gas, water, and energy must be identified and quantified for this stage of the process. If there is transportation of the product to and from an off-site maintenance facility, all transportation inputs must be included.

Reuse includes any on-site reuse and off-site reuse (Figure 8-5). On-site reuse may include intentional reuse of the product for the original purpose or incidental reuse for a different application. Examples include the use of an old coffee can to store screws, nuts, and bolts; old bottles to store old paint brushes; tires used as children's swings; and tire inner tubes used as water floats. A user may also reuse materials for an intended purpose. For example, the "reject" parts and scrap runners from Acme Plastics & Plating Company's

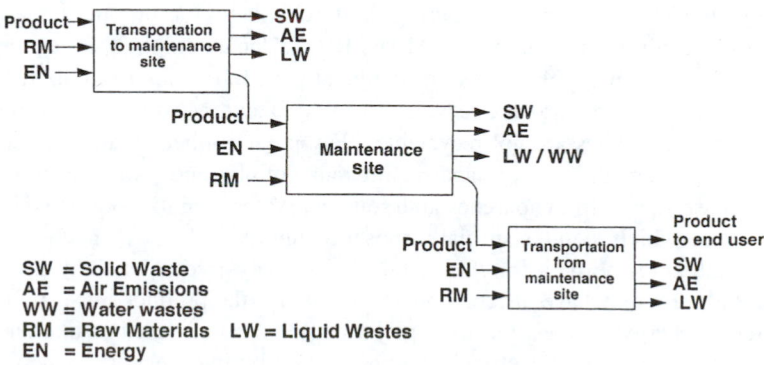

Figure 8-4. Off-site maintenance.

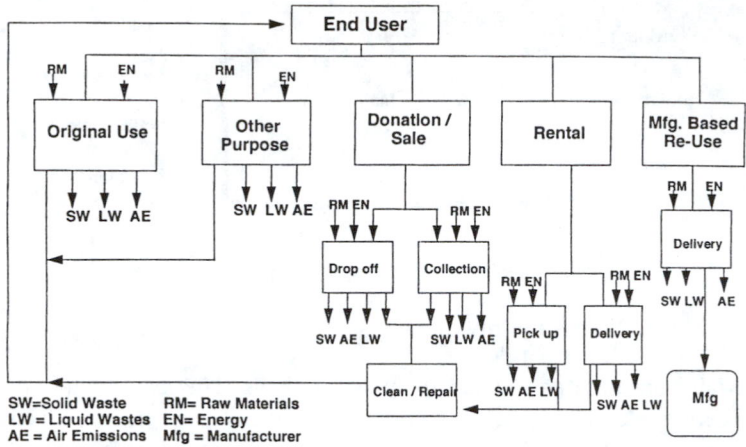

Figure 8-5.   Reuse.

injection molding process may be ground up, dried, mixed with virgin material, and reused in the injection molding process.

Reuse also includes reuse by another party for the product's intended use or return of the product to the manufacturer to be reused for its original purpose. Examples include refillable beverage bottles, computer keyboard components, and copier toner cartridges. The use by another user-party may entail the sale of the product by the original user or donation of the product. The product may be refurbished before reuse. Any energy or materials used to facilitate reuse must be included in this stage.

Discard involves the actual discarding of the product (Figure 8-6). The product will enter into either a waste management or a recycle system. At this point, the end user has taken the product out of service and will no longer use it. The waste management or recycle system will be addressed in further chapters.

The depth of the data acquired in this step is dependent on the intended use of the life cycle analysis. During the URM stage, the product may have a serious environmental impact or MANPRINT. The automobile, for example, uses nonrenewable fuel. It uses, for the most part, lubricants from nonrenewable sources. Spare parts are recycled to some extent. Most plastics currently used in automobiles are not recyclable. New paint applied to automobiles is not environmentally friendly and requires air pollution control measures that in turn use energy from nonrenewable sources. Water used for cooling contains antifreeze which requires special disposal techniques.

Gathering the data for this stage of the analysis can be very difficult. Similar users may have different environments for the product's use and have different energy sources. For example, a home computer used in one part of the country may have electrical energy supplied by hydroelectric generators; another may have nuclear, while another could easily have coal- or natural

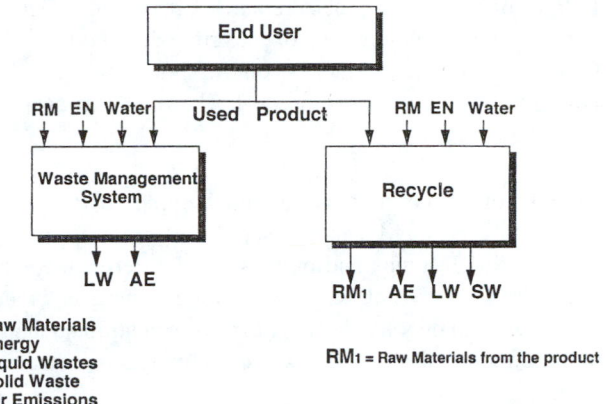

Figure 8-6. Discard.

gas-fired generators. Each of these sources has a different environmental impact. Breaking this down for an analysis for a computer company would be almost impossible. Making note of it would be acceptable. The analysis would therefore consider the amount of energy only and mention the potential sources. Because the data gathering is complex and critical, three basic approaches have been utilized for this stage:

- Primary data collection, where the end user directly describes the product is used and provides as much of the necessary data as possible.
- Secondary data, where published data such as article, studies, and surveys are used.
- Assumptions, where the analyst makes assumptions about the parameters of the product's use.

No single source is usually sufficient to conduct the life cycle inventory. In most cases, the data comes from a combination of the above sources. Each source has its own limitations, advantages, and disadvantages as discussed below.

PRIMARY DATA COLLECTION

Primary data collection can be done by statistical sampling. Here, the primary data collection is designed to capture a representative sample of data end users can provide on the parameters of the products use. For example,

- Other materials used in conjunction with the product's use and maintenance.
- Frequency of product repair or maintenance.

- Length of time the product remained in service. Details about its service, such as continuous or interrupted, operating conditions.
- Other uses for the product beyond its intentional use.
- How the product was disposed of when the end user was through with the product.

Frequently, the end user cannot directly supply information on inputs and outputs. The end user may be able to supply information on how the product was used from which inputs and outputs can be derived. Most of this information the end user may not be aware of unless the user is an institution or commercial user who may have some of the information such as energy inputs and air/water/solids emissions and wastes. This type of data collection takes several forms:

- Mail survey. A questionnaire is mailed to selected recipients.
- Panels of users that maintain logs or diaries of the products use.
- Personal interviews.
- Telephone survey.

Each form has its advantages and disadvantages. This type of data collection can be difficult and expensive. National surveys have been conducted in a statistically valid fashion with methodologies that are well defined and codified. Most of this type of data collection is done by independent survey research companies.

Primary data collected that is statistically significant have common phases:

- Experimental design and protocols — A description of the objectives of the research, population of interest, and information to determine the sample size, the source of the list to be used for sample selection, sampling frequency, survey time frame, means of handling nonresponse, and other information.
- Survey instrument — The questionnaire used to collect the data, including instructions to the interviewer.
- Data tabulation and extrapolation — The tabulation and presentation of the survey and means to show that the sample is representative. A description of the methods used to extrapolate from the studied population.

When the source of the data is statistical sampling, even if well designed, there are issues that include the following:

- Cost
- Representativeness of the sample

- Bias in the questionnaire
- Respondent bias
- Nonrespondent bias
- Statistical interpretation

There are other sources of primary data. These include a smaller sample size and may not be representative of the total population. However, for products that are used by a small segment of the population, are used on a regional basis, or are geared for a certain segment of industry, the following techniques may be employed.

- Focus groups — This is a group of 8 to 15 end users that are assembled in a room to answer questions about the product. This is usually conducted by professional marketing or survey companies.
- Suppliers of raw materials — For example, soap manufacturers for washing machines.
- Trade associations.
- Manufacturers of equipment associated with the product.

Issues associated with small groups or limited primary data collection are similar to those of the statistical sampling. The advantages include lower cost.

SECONDARY DATA SOURCES

Secondary data sources are any published or unpublished data articles, reports, or studies relating to the product that are available in published form. In many cases, good secondary data can be more comprehensible than limited primary data collection. Examples of secondary data sources include

- Trade, professional, or industry reports
- Articles in trade or professional journals
- Industry market survey and studies
- Federal government reports and statistical data documents and studies
- Reports of government rule-making hearings

The data from secondary sources may have been created for another purpose; therefore, some of the desired data may not be directly present. However, it may provide information on the expected or actual useful life of the product, maintenance of the product, and methodologies used to estimate the current data. Some government reports may include most of the data needed for this phase of the life cycle inventory.

Issues with secondary data include

- Cost — Usually less than primary data.
- Bias — Depending on the secondary source, the data could be biased in favor of the product or manufacturer or against it. Care must be taken to ascertain any bias in the data.
- Timeliness — Secondary data are usually compiled at a time prior to the current study. It is imperative that the data be representative of current practices.

ASSUMPTIONS

Assumptions are used when primary or secondary data are not available. Assumptions can be based on data that are not specific to the product or service being analyzed. For example, data on taking photos with a particular camera could be extrapolated to a single photo; washing a load of dishes in a dishwasher could be extrapolated to washing glasses. In most cases, the assumptions are used because data are not fully available.

When dealing with the end user, the patterns of user behavior should be included. The useful life of a product is not always included in an environmental life cycle analysis. However, to do an adequate job of assessing the MANPRINT of the product, the useful life should be included. The useful life of the product has three critical factors. The first is how the product was designed and built. This includes the materials of construction. The second is the environment and frequency that it is used and the third is how the end user perceives the product's usefulness.

The first determinate is built in by the manufacturer. If the product is used according to its intended use and maintained as specified, the product will have a life expectancy close to what the manufacturer states in the product literature. The second factor is more closely related to the user patterns. How well the end user follows directions and maintains the product can add or subtract useful life from the product. The third is basically how the user sees the product. If the product were tires on a front end drive car and the user didn't rotate the tires but otherwise kept them properly inflated, and didn't abuse them, it is not uncommon to get double the life out of the rear tires. If paper diapers are to be made to support a baby over a specific period of its life, the manufacturer would need to know how many times the "average" baby is changed per day during the time period of interest. In the case of detergents, if the manufacturer makes the detergent stronger so the user doesn't require as much to wash a load of wash, that should save the user costs. However, if the user doesn't change the way he or she thinks and acts about the product, there will be either no savings or it will actually cost the user more money to use what he or she thinks is the same product.

How the end user thinks about the product will have a major impact on its success. In some areas, the addition of certain environmental considerations can improve the product's chances in the market; an example is Body Shop's skin care products which are environmentally and animal-friendly products.

It should be noted that the obtaining of data and the sources for all the stages of the life cycle inventory is pretty much the same. The analyst will use primary or secondary data or assumptions. The treatment of the data should be similar to what is discussed above, no matter what stage the analysis is in.

9   RECYCLE — WASTE MANAGEMENT

The product has undergone most of its life cycle at this point. It has been manufactured, distributed, sold, used, and now it has reached the end of its expected life. In the past, the natural termination of the product would have been, most likely, a landfill if it were a solid object, the sewage treatment plant by way of a sewer if a liquid. The environmental movement started the country thinking about the rate the landfills were filling up and the amount of "stuff" that was being dumped in sewers. The effects of these practices were put under a magnifying glass. We did not like what we saw. For example, thousands of hair dryers are dumped into landfills every year. The main problem with most of them is that the on/off switch is broken. The whole hair dryer goes into the landfill. The landfill gets, on a daily basis, such things as paper wastes, yard wastes, glass, plastics, metals, etc. In some locations around the country, there are recycle programs for certain wastes. Typical recycle wastes are shown in Figure 9-1. The main reason any recycle program is successful is money. If it is cost effective to recycle a particular material, a higher percentage of it will be collected and recycled. The other form of financial incentive is fines for not recycling a material. Both incentive programs are based on money. If it costs more to collect and transport the materials than they are worth (i.e., it's cheaper to use virgin materials) then the materials will go into a dump.

Normally, people think of recycle and waste management as something that is done with paper, plastics, glass, and metals such as newspaper, plastic bottles, glass containers, and metals such as automobile bodies (auto recyclers). In the case of plastic bottles, the high-density polyethylene (HDPE) and polyethylene terephthalate (PET) plastics have a fairly well-established recycling infrastructure in the U.S. For the other recyclable thermoplastics, the infrastructure is somewhat weak. Up until recent times, corporations and institutions looked at recycling much the same way as the average American. Corporations and institutions "recycled" materials such as those shown in Figure 9-1. Recycling to most companies also includes the reuse of materials

Typical Recyclables in the Waste Stream

Office	Retail Store	Restaurant
ledger paper, white	ledger paper, white	cardboard
ledger paper, color	ledger paper, color	aluminum cans
computer paper	computer paper	ferrous metals
cardboard	cardboard	glass containers
newspaper	aluminum cans	polystyrene
mixed paper	glass containers	plastic bottles
aluminum cans	polystyrene	grease
glass containers	film plastic	newspaper
polystyrene	wood	computer paper
toner cartridges	unsold inventory	
obsolete office supplies	office supplies	

Schools	Industrial Facility	
ledger paper, white	Materials are specific to the type of industry, but may include:	
ledger paper, color		
computer paper		
cardboard	ledger paper, white	wood
newspaper	ledger paper, color	waste oil
mixed paper	computer paper	solvents
aluminum cans	cardboard	mylars
ferrous metals	newspaper	precious metals
glass containers	mixed paper	obsolete office supplies
polystyrene	aluminum cans	recoverable inventoty
plastic bottles	ferrous and nonferrous metals	polystyrene
wood	glass containers	film plastic

Figure 9-1.   Typical recyclables in the waste stream.

in-house during the manufacturing process. The concept of taking a used product back and recycling it was a foreign concept.

The next major stage in the life cycle inventory is "discard" (Figure 9-2). At this point, the intended end user has determined that the product is no longer suited for continued use as designed. There are multiple approaches to the discarding of the product. The major methods include but are not limited to:

- Donation/sale — If the product is still usable but no longer fits the end users requirements, the product can be sold as used equipment or donated to a charity, university, school, etc. Examples include old personal computers, industrial machines, copiers, etc.
- Recycle — The product is transferred to a recycle location for dismantling and salvage of useful materials and components. Sub-assemblies and components may be returned to the original manufacturer for cleaning/refurbishment and reuse in a new product. Examples include computers, electronic equipment, automobiles

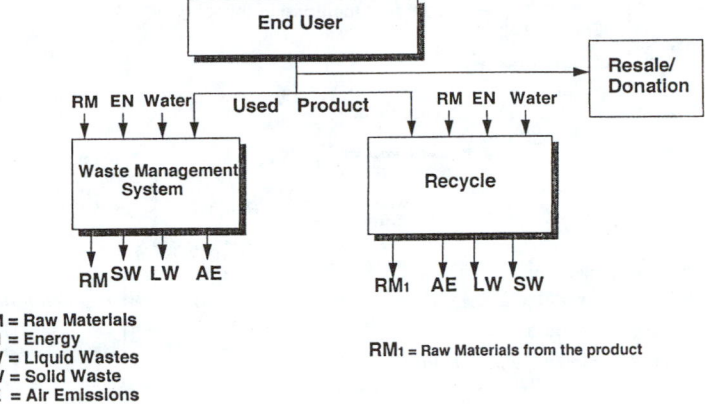

Figure 9-2.   Recycle/waste management.

and automotive components, scrap metals, "sludge" from plating wastes that contain certain heavy metals, copier and printer cartridges, plastic bottles, and scrap thermoplastics (see Figure 9-1).

- Waste management — The product is transferred to a waste management facility. This facility may be a sanitary landfill, hazardous materials landfill, incinerator, composting system, or other termination technology. However, in the past few years, waste management has added the waste minimization banner. In this mode, efforts are made to reduce waste before it is generated, reuse materials, recycle/recover materials, and ultimately dispose of by-products and nonrecoverable-nonrecyclable materials.

DONATION/SALE AND RECYCLE

In the discard stage of the inventory, careful consideration of the boundary conditions is required. If the analyst is not careful, this phase of the study can become extremely expensive. In each of the scenarios shown above, certain sets of data will be required for the life cycle inventory. The "usual" condition involves multiples of the steps shown above. The following discussion will provide some guidance in the conditions surrounding each of the above.

The donation/sale of the product to a third party after the primary user is finished with the product is a growing factor in electronics and industrial equipment. The end user can recover some of the initial investment (which he has already written off) or gain a tax advantage. This is also a boon for the recipient. The new owner probably could not afford new equipment, and this approach provides a winning situation for both parties. The analyst must decide where to draw the boundaries. If the boundary conditions stop at the point where the new user takes possession, the inventory for this stage is complete. However, if the requester of the analysis wants the total life cycle, the inventory

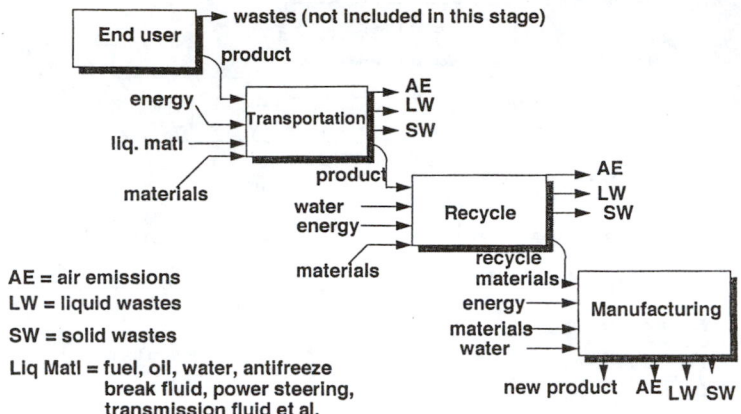

Figure 9-3.   Recycling.

will include a new end user and a secondary disposal activity downstream. In this case the techniques discussed in Chapters 8 and 9 will be repeated.

Recycled materials are those postconsumer materials and by-products that have served their intended use by a consumer (including businesses) and have been separated from solid wastes for the purpose of recycling (Figure 9-3). The two most prevalent types of recycling are the open and closed loop approaches. In the closed loop system (Figure 9-4), the product is recycled back into the same or similar products. In this scenario, the recycled materials have reduced the requirements for the production and disposal of by-products of virgin materials. This scenario does include the impacts of the recycle process (e.g., shredding, washing, melting, transportation) and any impacts incurred resulting from the use of recycled materials. The recycle process can run from the simple to very complex. An example of simple recycling is the return and reuse of glass milk bottles. A complex recycle can include the recycle of some metals, plastics, and recycle processes that are multiple steps. These processes, including all incoming attributes such as energy, materials and water, and outgoing items such a recycled materials, liquid wastes, air emissions, and solid wastes, must be accounted for.

The other scenario is open loop. In this case a product or parts of the product are recycled into alternative product or products (Figure 9-5). The overall change in the sequences of independent product A and product B is the introduction of recycled materials from product A. This has reduced or eliminated wastes from product A and reduced the requirements for virgin raw materials for product B. Figure 9-3 defines the basic data requirements for recycling, both open and closed loop types. The transfer step shown in Figure 9-3 should be handled in the same manner as the transportation steps earlier in the life cycle inventory. The basic reason for including the transportation step is to gain both a total and real picture of the MANPRINT and to assess the costs associated with this step. For example, Xerox, Canon, and others

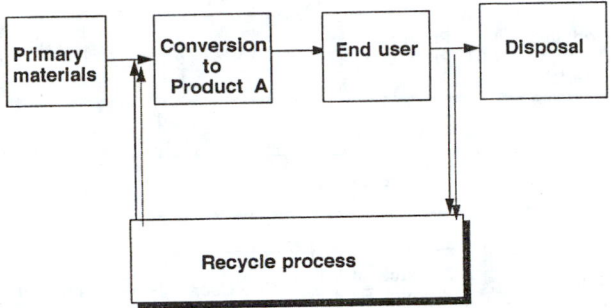

Recycle of materials/product to similar or same product

Figure 9-4A. Closed loop recycling. Recycle of materials/product to similar or same product.

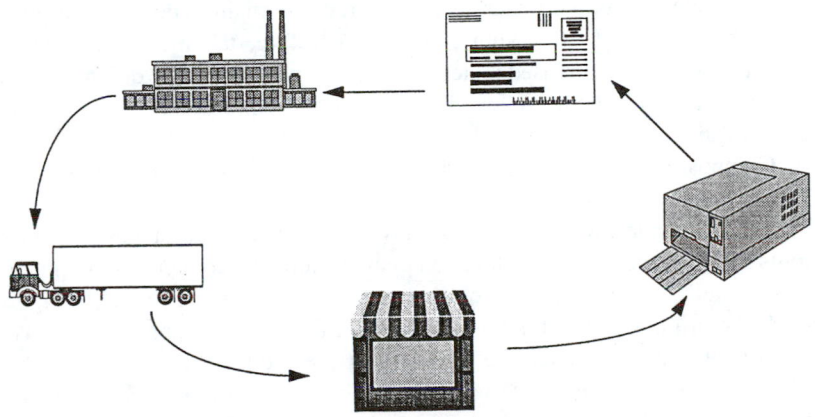

Figure 9-4B. Recycle of toner cartridges.

supply a prepaid mailer for the return of toner cartridges from their copiers, laser printers, and facsimile machines. 3M ships its bulk videotape in plastic foam packaging, then collects the used packaging at the customer's plants and reuses it. The Big Three automakers recycle used paint for use as filler in road repair and truck underbody paint.

Some products do not undergo open or closed loop recycling exclusively. The use of recycling is dependent on costs associated with the process, transportation, and the marketability of the recycled materials. Some recycled materials go into multiple products. Plastic bottles can be recycled into insulation, other bottles, carpet or clothing fabric, park benches, and plastic pallets. Some "new" products will be manufactured with varying amounts of multiple recycled materials. In general, recycled materials are mixed with virgin materials in the product. This is especially true in the case of molded plastics. The

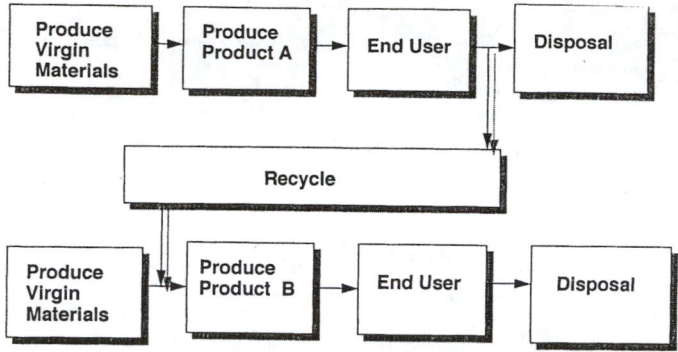

Recycled materials used in an alternative product

Figure 9-5.   Open loop recycling.

use of recycled or reused materials can degrade or change the physical properties of the product to varying degrees. Therefore, the molder will use a percentage of recycled/reused material with virgin material. By doing this, the molder reduces the demand for virgin materials while not endangering the quality of his product.

Composting is another form of recycling. It is usually considered a type of open loop recycling.

The life cycle inventory process has been divided into stages. The data inputs for the recycle and sale/donation phases should be collected as a separate stage as well. The basic input requirements are shown in the model in Figure 9-3. If the original product is donated or sold, the data on transportation, use, and ultimate disposal should be added. That the product has a "second" life is important to the analysis. The maintenance step in the second life of the product may be drastically different from that of it's "primary" life. The aging and wear of the product may require more frequent repairs. An example of this is the typical used car. The primary owner doesn't usually scrap the car when he or she is finished with it. The car is traded in on a new model and the "old" car is resold to a new owner. Data on the life cycle inventory for "second" owners may be available as averages based on the type of product involved. These figures were gathered and analyzed over long periods of time and, depending on the product, were considered fairly accurate for most life cycle inventories. This data collection is similar to the processes and models in Chapters 8 and 9. The inclusion of the data from the sold/donated stages is to be predetermined by the user of the life cycle analysis. All emissions and wastes from the transportation and recycling process are to be noted. The use and disposal of hazardous materials from a recycle process should be defined and accounted for. This can have a greater impact on the environment than if the product were simply disposed of by landfilling or incineration. All wastes from the recycling process should be noted and quantified.

As recycling technology improves in the coming years, it will complicate the models we have used. The newer technologies will allow products to be increasingly recycled in both open and closed loop fashion. This will present increased difficulty in the allocation of costs. Every effort to determine the costs split between recycled output is needed. In some cases, even though the output materials have multiple uses, they are produced by the recycler in one form, in the case of most plastics, as a powder or granular form. The next users (same or similar product or different product) obtain the material in one form. The total cost of recycling is determined, and the costs can be allocated based on sales to different customers.

For recycling, some of the same data sources used in the use/reuse and maintenance stages are applicable. Data on recycling activities may be more limited but many good primary and secondary data sources are available. The actual recycling companies themselves as well as professional organizations and the producers of the raw materials can supply a surprising amount of good-quality data. Government agencies and technical and trade journals are excellent sources. For products and materials in which recovery and recycling rates are low or recycling is relatively new, the data will be much more limited.

WASTE MANAGEMENT

This section describes the handling of wastes from the processes that have not been reused or directly recycled. It is important to note, early in the discussion, the difference between wastes and coproducts or secondary products. Coproducts and secondary products are products that are made concurrently with the principal product or they are made from the same or semiprocessed materials from the production of the principal product. The main differentiation of these coproducts or secondary products is that they have an intrinsic value or function to the company and to a customer. Wastes by definition are no longer of value and are to be disposed of. The models used for the life cycle inventory have indicated wastes after each stage and at the end for final disposal of the product. The treatment of waste management will be applicable to all the waste disposal throughout the model.

Products and coproducts must be clearly distinguished to assign the proper proportions of materials, water, wastes, and energy associated with their production. Life cycle inventories have typically used either the proportional amounts of the inputs and outputs (chemical equivalent or mass) or the relative economic value in proportioning the inputs and outputs. Most life cycle inventories use the proportional method. If an industry, particular company, or process has a high level or potential level of accidental releases, these will be added to the intentional releases.

For purposes of this text, waste management is defined as the methods used to handle, treat, and dispose of materials prior to their entering the environment. This includes waste minimization steps inside a facility, waste treatment (liquid or solids), or emission abatement systems (air) and landfills.

For purposes of the discussion, there are basically two types of waste: hazardous and nonhazardous. Hazardous waste examples include acids, strong bases, corrosives, and flammable and toxic materials including metals in solution, cyanides, radioactive wastes, incinerator wastes, mine tailings, biohazardous wastes, etc. Nonhazardous waste examples include most household wastes, paper, most solid metals, yard wastes, treated municipal wastewater treatment sludge, construction and demolition wastes, etc.

The environmental community has established and codified most states and in the U.S. Environmental Protection Agency (U.S. EPA) a hierarchy of preferences of waste management techniques. The hierarchy includes six preferences in descending order:

- Waste minimization/waste reduction
- Reuse
- Recycle/material recovery
- Composting
- Treatment (physical, thermal, chemical, or biological)
- Disposal (landfill, ocean dumping, groundwater or deep well)

Waste minimization and reuse are at the top of the list because of their importance to pollution prevention, resource conservation, and economic efficiency. Waste minimization reduces wastes by not creating them in the first place. Recycle-reuse minimizes the requirements for virgin raw materials. At this time, in the U.S., this occurs less frequently than waste minimization. The primary reasons for this are

- Management and technical people understand design and manufacturing better than recycle-reuse. Reducing the need for a material or reducing a material waste is "part of their job."
- There are intermediate steps such as collection, hauling, washing, grinding, sterilization, refining, etc., involved.
- Efficiency of recycle-reuse varies with the product, materials involved, distance to recycle-reuse locations from source, and the market for the materials.

Waste minimization and recycle-reuse reduce the environmental loading of a product but are, or can, be difficult to baseline. Recycle-reuse is easier to quantify because it could be measured. As for waste minimization, a baseline of where the company is at in a given point in time must be used. Reduced environmental loading at future dates can be measured relative to the established baseline.

The remaining waste management alternatives involve collection, processing, and ultimate disposal by or at a waste management facility. The inclusion of the waste management phase of a products life cycle is very important.

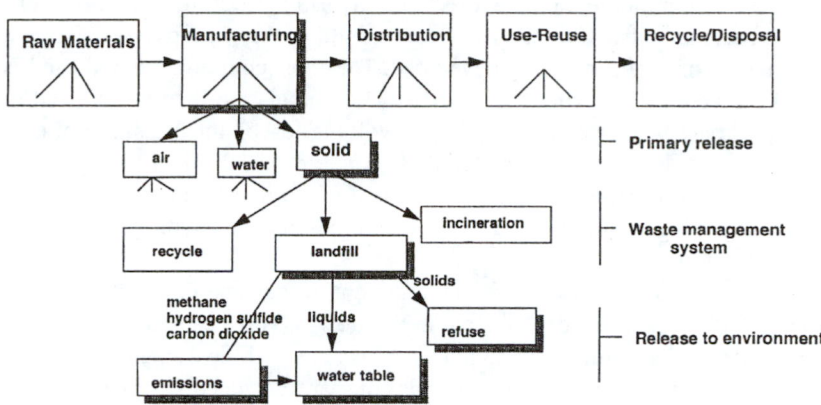

Figure 9-6. **Waste management/disposal. Generic pathways to the environment.**

This is the phase where the waste or spent product now reenters the environment. Up to this point, the product has caused certain wastes/emissions to enter the environment from its manufacture, transport, use, and maintenance. The product has now become obsolete, and the portions, if any, that can be recycled have been extracted. The whole or parts of the product remaining will now need to be disposed of. This is the final contribution to its MANPRINT on the environment. Figure 9-6 shows the generic relationship of waste management/disposal to not just the final stage but all the previous stages as well. The establishment of boundary conditions is important in this stage as well. The boundary of waste management system starts where the waste is generated. While there are, or may be, multiple waste management systems that can be employed, the analysis will only take into consideration whenever possible the actual waste management system used. For certain classes of products, this is straightforward. In other cases, the specific or multiple geographic location of the product's use will need to be factored in. For example, if the only viable disposal method for a product is a landfill, the analyst will focus on the contribution the product makes to the landfill and any "discharges" it may contribute to while in the landfill (methane, etc., from decomposition). In Figure 2-2 the volumes of 1000 plastic and paper bags from a grocery store were compared. The height of 1000 plastic bags stacked together was 4 inches. The same number of paper sacks was 46 inches. Obviously the paper has a larger volumetric impact. Neither one decomposes well in a sanitary landfill, so the volume is important. On the other hand, some locations may facilitate an incineration process as well. Materials have varying amounts of latent heat or energy stored in them. A comparison of internal energies for different products is shown in the Appendix. In this case, the product's emissions from both waste management processes will need to be considered. The technologies for waste management vary widely on a geographical basis.

There are multiple methods used for performing the life cycle inventory on disposal methods. The method selected should be the one most cost effective for the type of inventory being developed. The most common method used is based on weight or volume of materials. The outputs are then normalized to the inputs based on a weight-to-weight or volume-to-volume basis. It is almost impossible to do a complete materials balance, especially if a biological or some chemical process is involved. This is due to contributions not included in the incoming materials input data (examples include hydrogen, oxygen, enzymes, etc.).

Waste management includes the treatment, if any, and disposal of wastes from the life cycle inventory as noted in Figure 9-6. Three categories of wastes to be addressed are solid, liquid, and air emissions. All the treatment schemes that are used today to treat waste emissions are designed to transform the waste materials from their existing form to another form that is less toxic or has less impact on the environment and health of organisms around it. The real effectiveness of these treatments can be debated. It should be noted that one type of treatment for each generic type of waste will not work. For example, treating some industrial wastes, containing heavy metals such as copper, as domestic sewage will not only not work, it will prevent the sewage treatment facility from adequately treating the domestic sewage. Table 9-1 provides a general listing of types of waste management technologies. Most municipalities require industrial discharges that include materials that are hazardous to the sewage treatment plant to be pretreated to prevent upset of the sewage treatment facility. Some municipal treatment works also require industrial pretreatment for waste streams that will input a high organic load on the plant, thus requiring extra treatment capabilities. This is usually noted as BOD or COD levels (see Figure 9-7). Domestic sewage treatment technology has not changed much in the past 80 years. However, good secondary or

Table 9-1 Waste Management Technologies

Solid wastes	Liquid wastes	Air emissions
Landfilling (nonhazardous wastes)	Sewage treatment	Incineration/combustion
Landfilling (hazardous wastes)	Industrial treatments	Activated carbon
Incineration	Incineration	Condensation
Composting	Underground injection	Dust collectors
Landspreading	Lagooning	Scrubbers
Vitrification	Filtration	Filters
Fixation/cementation	Solidification	
Autoclaving/microwaving		
Recycling		
Illegal dumping		

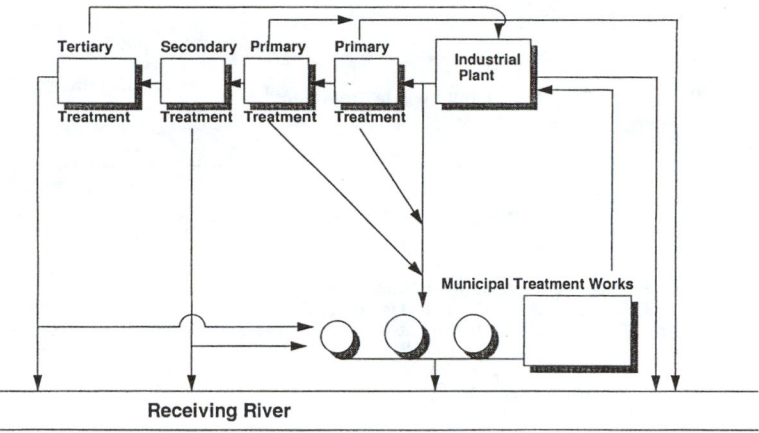

Multiple possible flows of Industrial Waste Treatment Systems

Figure 9-7.   Industrial/municipal treatment.

tertiary treatment of sewage does produce wastewater that can be disinfected and, in reasonable amounts at a time, be safely discharged to waterways. Industrial waste treatment has changed slightly in the past 50 years. Note again that different types of industrial waste require different treatment methods. There might also be more than one method for treating a particular industrial waste. The analyst must determine the method(s) used for the particular wastes for the product being investigated.

LIQUID WASTES

Liquid wastes in question need to be segregated by type and treatment methods required. The treatment method used will determine the required inputs and outputs. Most on-site treatments systems can be easily modeled with associated materials and costs from company records. For example, an on-site treatment system for plating wastes will have multiple waste streams for inputs with limited outputs (water, solid sludge; air emissions are included but may be extremely small). The inputs will vary, but will all be coming from the metal finishing operations at the plant. For example, chromate wastes and cyanide wastes will be from different plating lines. The analyst needs to determine what plating processes and rinses the product went through to allocate the contribution to the operation of the waste treatment plant. Off-site treatment is a different issue. In this case, wastes from the transportation method should be considered, unless it is insignificant compared to the actual treatment process. Off-site treatment processes need to be modeled to properly assign values to the input and output streams for the wastes from the specific product being studied (see Figure 9-8). The generic model block takes into

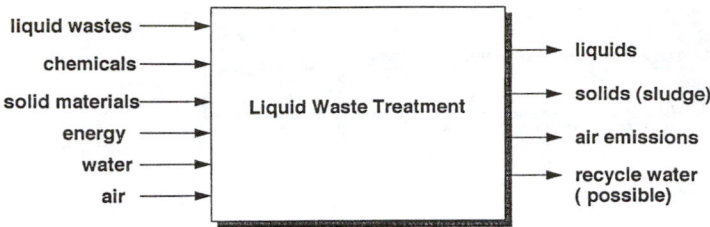

Notes: 1 Sludge may have commercial value
       2 Air emissions may be in the form of mists
       3 Solid material inputs may be thickeners, precipitation facilitators etc.

Figure 9-8. Liquid waste model.

consideration all the treatment systems used for water/liquid wastes. This includes but is not limited to the following examples:

- Reverse osmosis
- Precipitation/clarification
- Lagooning
- Filtration
- Underground injection
- Incineration
- Biotreatment
- Oxidation
- Land farming

The treatment of liquid wastes involves mainly in-solution treatments. However, the analyst should understand that treatment systems vary. Therefore, he or she should be familiar with the specific treatment system being employed. This will aid in determining the pathways for emissions from the treatment system to the environment. For example, a waste treatment facility might use aeration as a means of improving biological oxidation of the waste stream. The aeration process can and usually does release small amounts of the waste liquid into the air (environment) as well as some dissolved gases, volatile organic compounds (VOCs), aerosols, water vapor, and heat. The reader may, at one time or another, have had the misfortune of being downwind from a sewage treatment plant and been quickly able to determine his or her error in location. There are also the intended treated liquid effluents and usually solids in the form of sludge. Sometimes the liquid or solids from a treatment have commercial value. Sludges from a treatment system handling metal finishing wastes may have a high enough concentration of certain metals to qualify as an ore and be sold to refiners as raw material. In other cases the solids will find their way to a landfill.

Depending on the type of landfill, the analyst can usually count on the fact that the landfill leaks. Water-based seepage is trapped and treated on site in some landfills. Others seep into the ground or to nearby streams, lakes, or groundwater. Quantifying this seepage, if not captured and treated on site, will be difficult. The analyst should note that the condition exists or could exist. If there are data from public records at local agencies to help quantify it, the data should be gathered. The American Water Works Association, Water Pollution Control Federation, and other environmental organizations have valuable information on sources of wastewater. There are magazines and organizations that specialize in landfills as well. The data from these sources are reliable and as accurate as can be expected for generic settings. Local chapters of environmental organizations and Air Quality Management Districts may have quantifiable data on a specific local landfill. Local, regional wastewater control agencies, the U.S. EPA or the U.S. Army Corps of Engineers may have data on discharges to natural watersheds such a streams, rivers, lakes, and ocean outfalls. There are stringent requirements and permits needed for discharges to storm drains, wetlands, and waterways. These can be researched at the local offices of the local, state, and federal agencies. The requirements for determining the specifics for the "normal" inputs and outputs are obvious. The analyst should investigate leaks, spills, or overflows as well. These are potential pathways to the environment.

SOLID WASTE

The disposal of solid wastes is what people usually think of as garbage. This text uses solid disposal where most refer to land disposal. This is because land disposal is not the only method of treating solids that is available today. The disposal of solids includes

- Landfills
- Incineration
- Vitrification
- Alternative treatments (microwaving, etc.)
- Recycling
- Landfarming
- Composting
- Fixation/cementation

Landfills are the most common waste management practice. The other methods listed have niche applications. For example, composting is used primarily for natural organic wastes such as gardening and organic garbage. Incineration is used for liquid and solid materials that usually have some caloric value. Each has a set of inputs and outputs that will need to be noted and quantified most accurately as possible (see Figure 9-9). The emissions

Figure 9-9.   Solid waste model.

from an industrial incinerator are readily known by the operators of the equipment. This is because of the requirements of their permits. Typical emissions from incinerators are indicated in Figure 9-10A and B. Determining the allocation to a specific product, unless it is a captive system or the product constitutes a major part of its feedstock, will be difficult. The use of percentages will usually suffice. The other methods of waste management will offer varying degrees of challenges. For landfills, the following are considered the major outputs or pathways to the environment:

- Airborne emissions — Methane, VOCs, volatile liquid wastes, waste handling equipment emissions, landfill gas combustion products.
- Water emissions — Groundwater releases, leachate production and treatment processes, soil erosion, surface runoff.
- Solid particulate — Soil erosion, dust, sludge particles, chemicals applied to soils, compost, litter.
- Energy releases — Waste heat, landfill gas combustion.
- Fugitive emissions — Spills/accidental releases, containment system failures, waste handling equipment releases.

A similar set of pathways should be developed for the type(s) of waste management systems used for the disposal of the product or products being studied in the life cycle inventory.

AIRBORNE WASTE

Air emissions result from nearly every stage of the life cycle inventory. The emissions may or may not pass through some sort of waste treatment/control technology before reaching the environment. The concern of air emissions is the wider-spread impact air emissions can have on the environment. Solid-based wastes tend to stay in a relatively small area after discharge to the environment. Liquid wastes are capable of movement and can contaminate a wider area including groundwater. This movement can be fast (minutes) or extremely slow (years). Air pollution on the other hand cannot just spread over

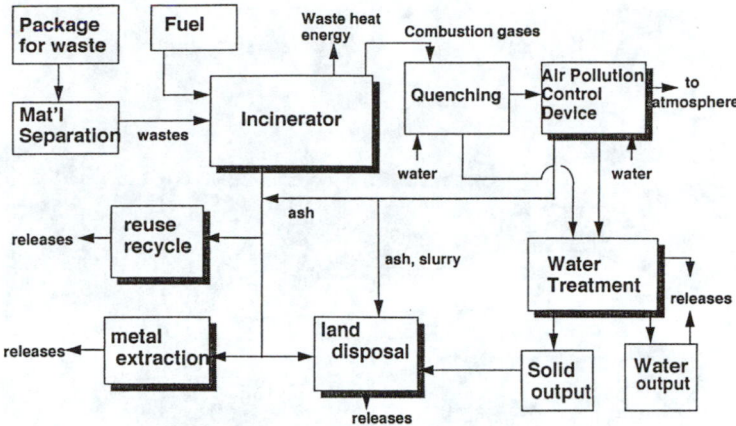

Figure 9-10A. Pathways for emissions from incinerators.

- **Emissions from incinerators include:**

    – carbon dioxide   - water vapor

    – water   - heavy metals

    – dioxins and furans   - waste heat

    – ash   - particulate organic chemicals

    – products of incomplete combustion

    – product specific pollutants

Figure 9-10B. Pathways for emissions from incinerators.

a vast area in a short period of time, it can contaminate soil and water, including both surface and groundwater, in a very short duration of time (see Figure 9-11). The boundaries used for airborne wastes and by-products need to be flexible. The extent of the inventory and associated costs need to be considered. To gain the most from an environmental life cycle inventory at a reasonable cost, the inventory should draw boundaries that include the major considerations and note the lesser ones. Air emission studies from waste management systems have been and continue to be studied. The U.S. EPA, air quality management districts, technical associations, and universities are good sources for quantitative information. The actual waste management organization may also be willing to provide information. Care must be taken with any data provided by a waste management facility. It is always a good practice to consider any such data as privileged and treat it as such. As with water and

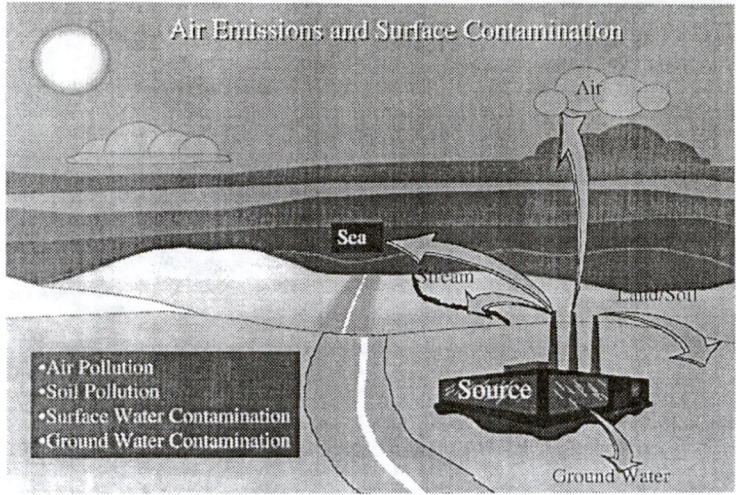

Figure 9-11.   Air emissions and surface contamination.

solid wastes, the units used to measure the inputs and outputs must be consistent. The units are usually in volume/volume or weight/weight relationships. It is confusing to mix volume and weight measurements. Therefore, it is necessary that the analyst

- Compiles a comprehensive list of emissions and pollutants from local, state, and federal laws, permits, and reporting requirements.
- Develops a standard measurement and reporting method.
- Develop consistent standards to fill in the gaps.

When the analyst is developing the data from air pollution control devices that involve a chemical transformation, care must be taken in rationalizing the data. For example, oxidation processes may *not* convert all or most of the product being treated to carbon dioxide. There are conditions that may exist that convert initial breakdown products into other compounds. These compounds may not be detected because no one is looking for them.

The collection and reporting of the data used for characterization of the waste generation and treatment will be a set of building blocks. Air, water, and solid wastes will be analyzed independently, recognizing that many processes will involve all three types of waste. It will also recognize that waste management processes may reduce wastes in one category only to raise wastes in another category. For example, look at a manufacturer, who, while trying to reduce airborne emissions of VOCs, changed from a solvent cleaning operation to aqueous cleaning. He reduced his airborne emissions but increased the amount of water he was using and increased the waterborne wastes emissions and increased his surcharges for BOD and COD.

Release	Amount	Comments
	Solid Wastes	
Hazardous		
Nonhazardous		
Municipal Waste	X kg	
Municipal sludge		other units or releases as neceaasry may be added
Compost		
	Airborne Wastes	
Regulated Particulate Toxic Carcinogenic Nitrogen oxides Sulfur oxides Carbon Dioxide VOCs Benzene other	Y Kg	other units or releases as neceaasry may be added
	Waterborne Wastes	
COD BOD TOC Suspended Solids Settleable Solids Carcinogens Heavy Metals Toxic Materials Benzene Sewage	Z Kg or applicable units	other units or releases as neceaasry may be added

Figure 9-12.   Total releases for the life cycle inventory component.

As shown in the fundamental modeling in Chapter 3, each phase of the product's life can be broken down into smaller steps until a fundamental understanding and model can be made. At this level the collection and quantifying of data including waste management can be handled easily. Once the data for each small step are generated (Figure 9-12), they can be aggregated up to the next level and so forth to finally produce the desired information at the required levels of the model.

The disposal model is the most complicated due to fact that wastes are generated all through the life cycle inventory model. The disposal section is treated like all the previous model steps with the addition of the wastes from all the prior steps. It is not uncommon to consider the wastes generated at each step as part of that step and use the models for the disposal step for the actions that take place at the end of the life cycle inventory. This approach simplifies the model and *makes the actual impact at each stage stand out*. The author recommends treating wastes generated at each stage at the particular stage where they are generated. Only the waste emissions generated at the disposal stage should be included at this stage level. Others may disagree. When it comes to methods to perform a life cycle analysis, there are no specifications regarding what must be included where in the study. If it is easier and more meaningful to include the wastes at each stage, then do it. If it is a minor consideration, due to the nature of the study, then place all the disposal/wastes in the disposal section. It is up to the analyst and the requirements and intended use of the life cycle analysis.

10   SUMMARY

The purpose of the life cycle inventory is to develop a life cycle assessment of a product. This assessment is chartered by the company management to aid in part of the product's design. The company may be interested in the impact it is having directly or indirectly through its products for any number of reasons. Most of the time it is due to some regulatory concern or to improve the company's image and customer market share. The life cycle analysis will provide an estimate of the impact the company is having on the environment or its MANPRINT. It will point out areas where the impact is the greatest and allow the user to compare his or her design or product to another competing design or product. It will not, however, directly determine what to do about any impacts. Waste minimization is another step. However, the life cycle analysis will provide guidance regarding where to put the emphasis on waste minimization and possible product/design improvements. It will also provide information to the design community to consider improvements during the design stage. That the designers cause most of the pollution from the product is not often clear to a design engineer. The use of the information from a life cycle analysis will help the designer better understand his or her role in a nonthreatening manner.

Knowing the various impacts the product has at each stage of its life will aid the company and its manufacturing management and the designers in determining what changes can be made and at which stage in the product's life cycle to reduce the product's MANPRINT. As mentioned in an earlier chapter, some copier and printer manufacturers have a no-cost return policy for their toner cartridges. This reduces the amount of refuse in landfills as well as the requirements for raw materials. The U.S. Army, during the early stages of a tactical missile program, instructed the missile contractors to do a MANPRINT study. They found that changing some materials and design callouts reduced the MANPRINT during manufacturing. They also studied what happens when they go into the field on training exercises. By making changes to fuel delivery systems, modifications missile support/launch vehicles, and troop operations, the Army could drastically reduce its MANPRINT on the environment (not counting the explosions when the missiles hit a target).

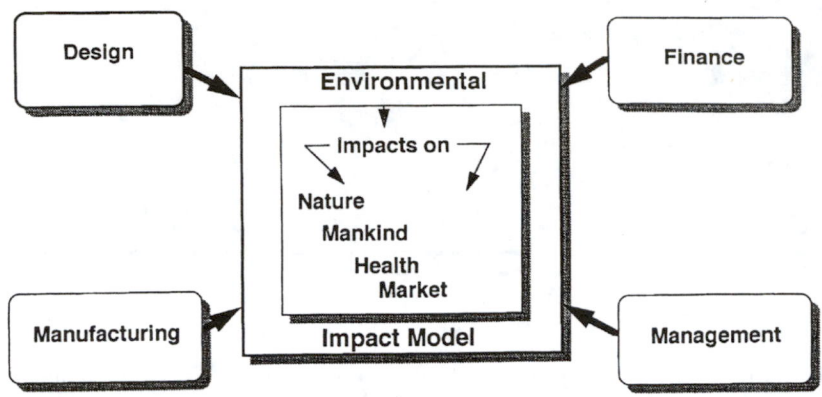

Figure 10-1. Impact assessment.

Sometimes it is satisfactory to provide the producer of the product(s) with information on the effects of its product on energy, raw materials utilized, disposal impacts such as volume in landfills, etc. Newer assessments have been taking a longer look at the assessment, more as a risk assessment.

At this point the life cycle inventory has become the life cycle analysis or environmental impact assessment. This is especially important to the management community. The impact analysis evaluates the environmental impacts caused by the product and its use/disposal. On the most basic level, resources are depleted and residuals are generated for each product and subproduct system. Resource depletion and wastes or residuals from the product and product manufacturing, distribution, use/maintenance, recycle/reuse, and disposal processes all contribute to the degradation of the ecosystem. Because of this, organizations such as design, manufacturing, finance, management, and marketing will be impacted by the results of the life cycle assessment. Figure 10-1 provides a simple diagram of this.

Environmental impacts can be organized into the following categories:

- Degradation of ecological systems
- Resource depletion
- Human welfare effects
- Human health effects

The type of models used to translate the life cycle inventory data to useful information will vary. The nature of the potential impact and the scope of the analysis will determine how much detail is required. For example, the release of chlorofluorocarbons (CFCs) into the atmosphere is reported to cause a degradation of the ozone in the upper atmosphere. This phenomena results in increased levels of ultraviolet light reaching the Earth's surface. This in turn is said to cause an increase in skin cancer. The models to show this are quite complex and the ability to ascertain the exact contribution of a company's CFCs contributed to the "problem" is almost impossible. In this case it would be sufficient to just state that the release of CFCs is contributing to the degradation of the ecosystem and possibly contributing to human health effects. It may be enough to simply estimate the impact on the company to eliminate CFCs. For a designer (product design or process design), the impact of the design on the environmental life cycle is one of the most challenging he or she will undertake. The following describes several aspects of environmental impact assessments on ecosystems.

DEGRADATION OF ECOLOGICAL SYSTEMS

Environmental life cycle assessments have largely focused on human health issues and to some extent on resource depletion. Up until recently, the degradation of the ecological systems has been largely ignored. This position is changing. This is because science has demonstrated a strong link to the health and welfare of humans to the health and welfare of the oceans, wetlands, forests, and other natural ecosystems. Mankind cannot as yet fully understand and recreate his ecosystem. This was demonstrated in the biosphere experiment. In this experiment, the oxygen levels in the biosphere went down to a level requiring outside ventilation. Some plants died. The reason was the cement used in the walls interfered with the carbon cycle, removing carbon dioxide that the plants needed to produce oxygen for the air and carbon for their cells. Bacteria created the carbon dioxide but the concrete absorbed enough to upset the cycle. It has been shown that if the oceans die, so does man. Therefore, governments and scientists are now starting to understand the fragile relationship man has with his ecosystem.

Ecological risk or impact assessment is modeled after human health risk assessments but is more complex. Each ecosystem the product will come in contact with will be evaluated based on the impact the product has on that system. Ecosystems need to be bounded as well. The whole world is an ecosystem; however, this approach may be too vast in most applications. In the case of CFCs, mentioned earlier, the atmosphere around the world was considered an ecosystem and the impact of CFCs on it was studied. Subecosystems, mainly human interaction, was added as part of the risk assessment.

Ecosystems are never static. They are living systems and as such tend to respond to stimuli or stress. Stress in the ecosystem world can be chemical (toxic waste dumping, pesticides, etc.), physical, (habitat destruction due to construction, logging, farming, etc.), or biological (introduction of a foreign or exotic species) in nature. These stimuli or stresses cause the ecosystem to respond in some manner. The risk involved to the ecosystem will elicit a distinct response. The response may range from a mild adjustment to a destruction of the existing ecosystem and the evolution of a new one. The risks to an ecosystem are classified as follows:

- Type of ecological response based on indicator's response
- Intensity of the potential effect
- Time scale for recovery after stimuli removal
- Spatial scale (local, regional, global)
- Transport media

The rate of recovery of an ecosystem from a stress is important. In some cases the effect is permanent changes in the biocommunity and species extinction. Stresses tend to be cumulative, finally resulting in large-scale changes and problems. Habitat destruction and atmospheric changes are widespread examples. Care must be exercised in this area. The use of renewable resources may be misleading, e.g., the use of wood to make paper and other wood products. The first thought and the advertisements from the lumber industry indicate that the "forests" are replaceable. The wood that the lumber mills want is, to some extent, replaceable. It can be grown on tree farms. These tree farms are *not* the same ecosystems that were placed here by nature. They do not harbor the same plants and animals that the original forest, that was cut down, did. Is this the same thing as the original? No. Is the wood replaceable? Yes. Is there an impact to the environment, Yes. Should this be taken into consideration during an environmental impact analysis? Yes!

Determining the potential risks and likely effects is the first step in the ecological assessment. The data have been provided by the life cycle assessment. Now the user must evaluate the data to determine the possible effects the product's design and life cycle will have on the environment and what, if any, changes can be made to reduce, mitigate, or eliminate the potential environmental impact.

RESOURCE DEPLETION

The extraction and consumption of natural resources is divided into two categories. The first one is renewable resources such as hydroelectric and wind-generated power and certain biomasses such as corn and other crops. The second category is nonrenewable resources. There are finite amounts of minerals in the earth. Oil, coal, iron, copper, chrome, and in some areas water, etc., are not being "remade" at any noticeable rate. Certain minerals are obtained only in specific locations in the world. Because of this, wars have and will be fought. Up until recently, most people considered minerals to be almost inexhaustible. People have had the "frontier" mentality of "rip it up, use it, and throw it away — we can get more" philosophy. The problem is we do not have infinite amounts of resources. Fresh water is becoming scarce because of overuse, leaching of minerals and pesticides, and the disposal of sewage and industrial wastes. The producer of products needs to understand what the demands on the environment and natural resources are and how to prolong their life. The life cycle analysis will provide the understanding that, with the present design of the product, the company will require "X" thousands of pounds of thermoplastic, "Y" thousands of pounds of steel screws, nuts, bolts, washers, "Z" gallons of water, etc. Now, armed with those data, the design and manufacturing engineers can work on how to reduce the requirements. This may involve the reuse of waste plastic, recycle of solvents, redesigns to reduce components parts, and more efficient or reduced use of water, thus reducing the number and types of fasteners such as nuts, bolts, washers, resources, etc. This can have major cost reduction implications. Examples include

Company	Operating change	Benefit
3M	Developed an adhesive for sealing of boxes that didn't require solvent	Eliminated the need for $2 million in pollution prevention equipment
Reynolds Metals	Replaced solvent-based ink with water based in packaging plants	Cut emissions by 65% and saved $30 million in pollution control equipment
Hughes Aircraft	Cut off unnecessary air sparging to a plating line	Reduced chemical costs, improved quality of the plating, and saved over $50,000 in pollution abatement equipment
	Recycle of materials and materials reduction including just-in-time chemical acquisition and use	Over $ 1.5 million saved Floor space saved >17,000 tons of materials recycled

The engineers can design the product for ease of consumer recycling of critical and expendable components such as toner cartridges. The designers can think of recycling and reuse as "mining" the old product for materials to reuse in the same design product or another product. *Mining the scrap* can be as profitable as mining for the natural resource. The designers may require some initial training in design for assembly (DFA)/design for disassembly (DFD-A) and design for the environment (DFE). They are all very closely related. DFE includes DFD-A. If the recycler or customer can't *easily* get to the components that are to be recycled/reused, it isn't cost effective and, if it isn't easy, the recycler/customer won't do it.

As noted above, the depletion of natural forests and the substitution of man-made ones is a resource destruction. Students at California State Polytechnic University at Pomona noticed this during their study of the comparison of plastic grocery bags to paper ones. After their study, plastic won out based on multiple environmental impact analyses.

HUMAN HEALTH AND WELFARE ISSUES

The one thing most manufacturers do not want is to have their product cause human health problems. Most do not want their products to harm animals either. Having a product unintentionally cause sickness, injury, or death to man or animal can have devastating effects on the company. Beyond the "normal" precautions taken during the design phase of the life cycle, the assessment can determine any long-term effects the product may cause. The cause may be indirect. The design may cause the manufacturing organization to use solvents, solutions, or materials that are hazardous to humans. The effects on humans is dose related but the requirements for the materials is caused by the designers. The effects of the product on humans and the local ecosystems is determined by the life cycle analysis. This will include not just direct effects but indirect.

Another aspect of human health and welfare is the loss of the natural environment, e.g., loss of forests, wilderness areas, wildlife, streams, etc. Most people don't understand nature — some appreciate it more than others — but the natural environment is not man-made greenbelts. When our products and activities cause the destruction and loss of natural environments, we all lose.

The use of a life cycle analysis to ascertain the impact a product or company has on the environment is a noble undertaking. The company using the tool may find out facts that it does not want to hear. Using the data to improve the environmental impact will benefit not just the company but the world outside as well. Companies that have taken the reduction of wastes and MANPRINT seriously have had some or all of the following benefits:

- Improved market share
- Reduced costs
- Additional advertising material

The use of the data must begin at the top of the organization. Management must be committed to reducing their company's impact on the environment. The design organization must understand management's commitment and use the life cycle analysis as a guide for reducing environmental impact through the design of the product. In most cases, training can be beneficial.

The training should be from companies that understand not just environmental issues (environmental rules, waste minimization, assessments, and life cycle analysis), but materials and processes (M&P), design (including DFA/DFE/DFD-A), and manufacturing. Most environmental consultants do not understand product design or manufacturing. Consulting or training firms that have the capabilities listed above to conduct this type of training are not easy to find (Figure 10-2). They must have people that have experience in all three areas. The following companies have the background to perform the necessary training and consulting required:

Company			
Name of consulting company:			
Address:			
Telephone/Fax. number:			
Contact name:			

Skill Summary			
Skill matrix:	Yes	No	Comments
Industrial design experience			
Direct manufacturing experience			
Environmental areas of experience			
Waste minimization			
Lifecycle analysis			
Remediation			
Site assessments			
Envir. management assessments			
Teaching/ training experience			
Waste treatment system design			
Publication/patents in fields			
Estimated costs			

Figure 10-2.   Consultant-training checklist.

Orion Management Company
Principal offices at
24000 Alicia Parkway # 17
Box 211
Mission Viejo, CA 92691

Edward C. Lai & Associates
Principal offices at
2424 229th Street
Torrance, CA 90501

Coopers & Lybrand
Offices throughout the U.S.

Arthur D. Little
Offices throughout the U.S.

The use of the life cycle analysis can range from a directed product/process or potential impact area to a full life cycle analysis starting from a figment of the designer's imagination through ultimate disposal. The type of analysis chosen will depend on the intended use of the study and the amount of time and money the company wants to spend in it. There is no one-and-only way to conduct a life cycle audit. Using the models shown earlier as templates, the analysis can be varied to accommodate the needs of the company.

The data from the models must be modified into a format or sets of formats that turn the data into information usable by the company. The use of charts and graphs will aid in the understanding of the technical data. The user wants information that can be readily used without laborious calculations and manipulations to translate into something usable in the designs. Figures 10-3 through 10-6 illustrate the use of graphic formats as part of the final report.

Each life cycle inventory has its own set of limitations. These limitations must be contained in the report and study notes. It must contain statements on components of the study that were not quantified or omitted and for what reasons, the base year of the study, special conventions used, data ranges, efficiencies, whether the study was peer reviewed, if reviewer comments were included, and any other limitations. Below are some examples of possible limitations that would be documented as such:

- How deep and broad was the study? How were the boundaries drawn?
- Are there any significant limitations that affect the data?
- Do facilities and processes meet minimum regulatory requirements?
- Sources of generated information. Type of data being used (primary, secondary, assumptions, tertiary). Statements as to reliability of the data.
- How did the analysis address any missing data?
- How does, or did, changing technology affect the study?
- How can the data best be delivered to intended users?
- How much proprietary information was used and how much can be made available to the intended users of the report? Was proprietary

PROCESS	ENERGY	ENVIRONMENTAL IMPACT	MANUFACTUREABILITY	RECYCLABILITY
Paper	• 1,340,000 megajoules per bag required • Numerous equipment transfers to complete a finished product • Paper uses approximatley 2 times the energy than it takes to produce plastic used for grocery sacks	• Odor producing process • Produces air and water pollutants • Inorganic fillers such as adhesives and inks, produce large quantities of wastes • Deforestation and loss of ecosystems • Carcinogenic products produced • Limited biodegradability • Large amounts of solid wastes	• Messy manufacturing operations • Numerous operations to finished product • Government regulations have caused a large increase in capital improvements to control pollutants • Increased off shore production where environmental laws are less stringent or can be overlooked	• Low profit margin in recycling paper bags • Limited degree of reprocessing • Secondary fibers are of a lower quality • Increased solid wastes • Very high capital investment

Figure 10-3. Data summary sheet.

Process	Energy	Environmental Impact	Manufactureability	Recyclability
Plastic	• 580,000 megajoules per bag required • Can be incinerated has latent heat content • Uses approximately 1/2 the energy required than it takes to produce paper sacks • Takes less energy to recycle	• Thermal decomposition produces aldehydes • Water pollutants; waster wastes through production and polymerization • Toxic additives used in processing facilities • Littering of post consumer plastic; trash in streets, oceans and wilderness areas • Take much less volume in a landfill than paper sacks	• Inert final product • Low energy consumption • Gas blown • Polyethylene pellets used • Water based ink • Uses additives such as oxidizing agents/ colorants	• Loses desired properties • Contamination losses • Chemical additives • Additional additives required for biogradability or photodegradability • Easier to recycle than paper sacks • Less pollutants recycling plastic bags than paper

Figure 10-3. *Continued.*

Wt	Attribute	Plastic	Paper
5	Environmental Impact	8	6
4	Recyclability	7	5
3	Generated Waste Mtrl	8	5
2	Energy	9	5
1	Manufacturability	7	6
	TOTAL (ATT X SCORE)	117	81

Figure 10-4. Table comparing key attributes for plastic and paper.

information manipulated to avoid the exact proprietary nature thus allowing general use of the data?
- What trade-offs were made in selecting the boundaries for the study?

The final report should include, but is not limited to, the following elements:

1.0 Executive summary
    1.1 An opening paragraph describing the nature of the study
    1.2 A brief summary of the results
2.0 Introduction
    2.1 A description of the purpose of the study
    2.2 A description of the products the study included
    2.3 The dates over which the study was undertaken
    2.4 The names of companies, institutions, government agencies, and individuals who were contacted during the study
    2.5 Limitations of the study

**The Environmental burden per
1 million paper bags**

Subject	Polyethylene bags	Paper bags
Energy required	580,000 megajoules	1,340,000 megajoules
Air pollution produced (kilograms)		
Sulfur dioxide	198	388
Nitrogen oxides	136	204
Hydrocarbons	76	24
Carbon monoxide	20	60
Dust	10	64
Waste water (kilograms)	10	512

Figure 10-5.   The energy dilemma. The environmental burden per 1 million bags.

2.6 Boundaries and assumptions used during the study

2.7 A summary of the environmental impact analysis

3.0 Stages of the inventory

    3.1 A description of each stage and sublevels in each stage

    3.2 Boundary conditions imposed and qualifications

    3.3 A description and *quantification* of inputs and outputs for each stage at the top level for each stage; describe anything

	Kraft		
Item	Unbleached	Bleached	Groundwood
Virgin material (dry)	1.0 ton	1.1 ton	1.1
Waste water Volume	24 x 10^3 gal	47x 10^3 gal	10x 10^3 gal
Energy	16 x 10^6 BTU	22 x 10^6 BTU	18x 10^6 BTU
Solid wastes	135 lbs	225 lbs	164 lbs
Air emissions	50 lbs	64 lbs	65 lbs
Waterborne wastes *	46 lbs	90 lbs	25 lbs

* includes BOD

Figure 10-6.   Environmental impacts of manufacture on one dry ton of wood pulp.

outstanding at each stage (outstanding = anyobvious issue or opportunity)

3.4 Procedural activities, limitations, assumptions

3.5 Summary of sources for each stage

4.0 Top level or product level summary or product/design comparisons

4.1 Discuss rollup of the environmental impact and comparisons to other similar designs/products if any exist

4.2 Discuss limitations and assumptions

4.3. Highlights of where the heaviest environmental impacts are located, causes (if known)

4.4 Comparisons to other products or processes that are pertinent

4.5 Conclusions

4.6 Recommendations

Appendix

Holds major tables, drawings, references, and support information

The final report is provided during the exit briefing to management and others that have interest in the analysis. The exit briefing provides for the presentation of the pertinent information and delivery of the final report. The exit briefing is an excellent time for management to reestablish the importance of the report and define near and long-term activities. Management support for future activities must be made clear if the life cycle analysis findings are to be utilized.

For results from the life cycle analysis to have a long-term impact, employees and management may have to change the way they think and act. This will not happen by itself. To facilitate changes throughout the organization, the individuals will need to understand what is going to be required and why. By addressing the unasked question "What's in this for me?", the company will not get the total expected results. To accomplish this, training will be necessary (Figure 10-7). Most of the training, for the organization, will be common. However, each department will have unique training that is required for their specific function. For example, accounting may be figuring out how to easily account for environmental costs above what is presently reported. This will allow management to easily review the TOTAL environmental costs. The collection of scrap and rework costs may be part of their training. The design engineers will get different training such as how to set up and use a "green design system," materials training, how design impacts the environment, and a crash course in manufacturing technology. Manufacturing will be concentrating on pollution prevention, tooling/packaging, etc.

The life cycle analysis and report is not the end of the process. If the study and report are to be of value, the report and subsequent findings must be used to reduce the environmental impact of the product or processes that were studied. Two areas that will have major impacts will be the design engineering organization and manufacturing.

Figure 10-7.   Training for life cycle analysis.

The design engineering function determines not just what the product looks like and how it works but what the materials of construction are and to a large extent what manufacturing processes are used. The design engineering function therefore determines the environmental destiny of the product. Aspects of the design determine how it is packaged for shipment and the amount of energy and materials required for use, maintenance, and repair. The final destiny of the product upon disposal is also preordained by the design engineers. The design community must consider the following for each product:

- Cost of the product
- Legal aspects of the design and product use
- Cultural aspects (the product's name, color, shape, etc.)
- Performance of the product
- Materials and processes for production
- Environmental destiny

Obviously, all the items listed above are interrelated. The life cycle analysis study, however, allows the designer to see where in its life cycle the product has the greatest impacts. This can provide the designer with some ideas as to what aspects of the design could possibly be changed to reduce the impact. Because the designer knows, by means of the study, how the product impacts

		Raw Materials	Mfg/Processes	Distribution	Use/Reuse	Recycle	Disposal
Product	inputs						
	outputs						
Processes/ Mfg.	inputs						
	outputs						
Distribution	inputs						
	outputs						
Waste Management	inputs						
	outputs						
Use/Reuse	inputs						
	outputs						

Figure 10-8.   Conceptual matrix.

the environment at each stage, he or she can more easily estimate the impact the proposed change will have on the environment and the product. A conceptual requirement matrix is shown in Figure 10-8. Using the knowledge of impacts the designer can focus his or her energy at reducing the environmental effects. The designer must, however, consider the functionality and cost of the product. If making the requisite changes drastically increases the cost, what will the impact be to the end user? Will the user go someplace else for a similar product? Sometimes the company can actually charge more for an environmentally sound product or gain increased market share. The marketing department needs to understand the changes to better sell the product. The changes to reduce environmental impact may start in the factory. The materials of construction, coatings, and cleaners required are predetermined by the designers. The company may want to institute a program of design reviews with manufacturing people or have designers spend time in manufacturing actually seeing, firsthand, how the product is actually built and what their callouts on drawings actually do to production. In most cases, the design engineers do *not* actually know the details of how the product is manufactured. What they think and what really happens are usually quite different.

Manufacturing engineers and management must also review the life cycle analysis. Manufacturing creates many materials that cause an environmental impact for the plant site. The impacts come from

- Solid waste such as paper, wood, plastics, metals, etc.
- Sludge from water treatment facilities for such operations as painting and plating lines. These can have high concentrations of heavy metals or paint sludge.

- Liquid wastes such as solvents, hot water, cooling water, process water, cooling agents, wash water, etc.
- Air emissions such as solvents from coating operations, vapor degreasers, lamination processes, welding/brazing processes, and microelectronics fabrication procedures.

Manufacturing operations in a company usually handle most of the chemicals used. Therefore, they create the largest amount of hazardous waste as well as nonhazardous wastes. Manufacturing can and should undertake a waste minimization program that includes

- Reducing the amount of chemicals used
- Changing to more environmentally friendly chemicals or processes whenever possible
- Reducing solid wastes
- Recycling/reusing as much material as possible
- Working with design engineering to reduce the requirements for environmentally unfriendly materials and processes

There are books on the market detailing waste minimization techniques:

- *Hazardous Waste Minimization Handbook* by T. E. Higgins (Lewis Publishers, 1989).
- *Waste Minimization as a Strategic Weapon* by D. F. Ciambrone (CRC/Lewis Publishers, 1995).

The largest single factor in the successful use of a waste minimization program as well as the life cycle analysis is *management*. Management must be the driving force behind the use of the life cycle analysis. Management will drive engineering and manufacturing to do their respective jobs. However, management must include marketing and finance as well. The evaluation of the position of a product with respect to its future and the amount of time and money that will be spent improving its environmental impact are management, finance, and marketing decisions. Figure 10-9 illustrates a matrix method of determining what to do with a product. It takes into consideration real and perceived environmental impacts. The real impacts can be determined from the life cycle analysis. The perceived impacts are determined by marketing listening to the consumer. Each product will position itself somewhere in the matrix. From its respective position, management must decide what strategy and leadership is required. A perceived impact is just as real to the user as an actual impact.

If the position of the product is in the high-perceived but low-actual impact area, management will be looking at marketing and communication strategies.

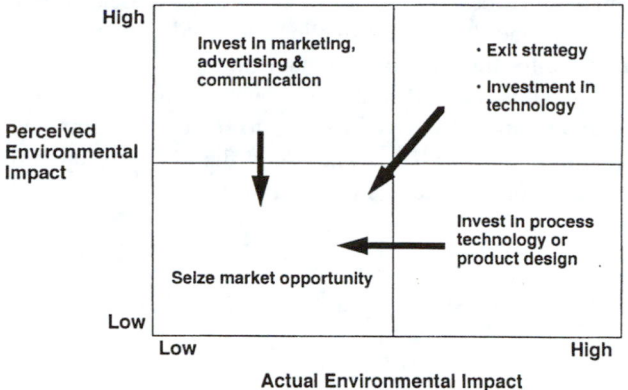

Figure 10-9.   Actual vs. perceived environmental impact.

If the product is in the high-actual impact but low-perceived impact area, the management will be considering process or product improvements. This will be an investment in engineering or manufacturing.

The high-perceived and actual-impact area may mean that management will have to invest in processes or product changes as well as new marketing approaches. The other alternative is to exit the product market. There might be cases where the product is in this block area and the market is still strong. Management may consider possible changes to the next-generation product and leave the present one alone.

For the product that falls into the low-perceived and low-actual impact area, the strategy should be to seize the advantage and try to improve market share by advertising the product's environmentally friendly nature, especially if the competing products by other companies are not as environmentally friendly as yours.

More companies are increasingly recognizing that, to achieve an environmentally competitive advantage, communication is necessary. Employees are possibly the single most important group to communicate to. Employees, if communicated to, will

- Strive to improve a successful situation. If you are doing well in reducing environmental impacts and tell employees how their efforts are helping insure the company's and their future, they will strive to make more improvements. Everyone wants to be part of a winning situation.
- Tell others about their company's product's low environmental impact. This is more believable to outsiders than an advertisement from marketing.

The company needs to tell customers, distributors, local community leaders, shareholders, and the general public. There will be opportunities during the product's life for this kind of communication.

Environmental issues will continue to be important to corporation's strategies and opportunities. Management has to direct the company through a myriad of issues on a daily basis. The use of the life cycle analysis can help management position its products for the future.

REFERENCES

1. World Wildlife Fund and The Conservation Foundation, *Product Life Assessments: Policy Issues and Implications, Summary of a Forum*, Washington, D.C., August 1990.
2. U.S. Environmental Protection Agency, *Information Resources Directory — Fall 1989*, Springfield, VA, U.S. Dept. of Commerce, National Technical Information Service, October 1989.
3. Hunt, R. G., Sellers, J. D., and Franklin, W. E., *Resources and Environmental Profile Analysis: A Life Cycle Environmental Assessment for Products and Procedures*, Environmental Impact Assessment Review, Elsevier Science, New York, Spring 1992.
4. Battelle and Franklin Associates, *Life Cycle Assessment; Inventory Guidelines and Principles*, U.S. EPA Risk Reduction Engineering Laboratory, Office of Research & Development, Cincinnati, OH, EPA/600/R 92-086.
5. Stoop, J., Scenarios in the design process, *Appl. Ergon.*, 21(4), 304, 1990.
6. Nemerow, N. L. and Dasgupta, A. D., *Industrial and Hazardous Waste Treatment*, Van Nostrand Reinhold, New York, 1991.
7. Ciambrone, D. F., *Waste Minimization as a Strategic Weapon*, CRC Press, Boca Raton, FL, 1995.
8. Boothroyd, G. and Dewhurst, P., *Product Design for Assembly*, Boothroyd and Dewhurst, Inc., Wakefield, RI, 1989.
9. Sellers, V. R. and Sellers, J. D., *Comprehensive Energy and Environmental Impacts for Soft Drink Delivery Systems*, Franklin Associates, Prairie Village, KS.
10. Little, A. D., *Disposable versus Reuseable Diapers: Health, Environmental and Economic Comparisons*, Arthur D. Little, Inc., Cambridge, MA, 1990.
11. Fabrycky, W. J., Designing for the life cycle, *Mech. Eng.*, 109(1), 72, 1987.
12. Saaty, T. L., *The Analytical Hierachy Process*, McGraw-Hill, New York, 1980.
13. SETAC, *A Technical Framework for Life Cycle Assessment*, The Society for Environmental Toxicology and Chemistry, Washington, D.C., 1991.
14. Little, A. D., Identifying Strategic Environmental Opportunities: A Life Cycle Approach, *Prism*, Arthur D. Little, Cambridge, MA, 1991.
15. Meadows et al., *The Limits to Growth*, Universe Books, New York, 1972.
16. *A Blueprint for Survival*, Club of Rome, Universe Books, New York, 1972.
17. Higgins, T. E., *Hazardous Waste Minimization Handbook*, Lewis Publishers, Boca Raton, FL, 1989.

GLOSSARY

Biodegradable   Capable of being broken down by natural, biological processes

Compatible Material   When combined compatible materials do not cause unacceptable impacts

Composting   An aerobic biological process employed to breakdown natural organic materials and organic garbage into compost for use in agriculture

Cross Disciplinary Team   A team consisting of representation from all affected operations

Concurrent Design   A design approach that causes the concerns and ideas from production and engineering to be considered at the same time during the design phase of a program

Downcycle   The recycling of a material or product for a less-demanding use

Embodied Energy   Energy contained in a material that can be recovered for use through combustion or other means

Life Cycle Analysis   Draw on the information from the life cycle inventory and identify issues that are pertinent to the particular process or products under consideration

Life Cycle Inventory   The portion of the life cycle study process that gathers the specific data on each of the product's stages and quantifies them

Impact Analysis   Assesses the environmental impact and risks associated with various products or activities; an impact analysis interprets the data from an impact inventory

Improvement Analysis   The strategic evaluation of the options for reducing environmental impact of the product and the initiation of actions

Inventory Analysis   Identifies and quantifies all inputs and outputs associated with a product and processes including materials, energy, and residuals

MANPRINT   The impact on the environment caused by humans and their equipment/products and processes

Physical Life Cycle   The series of physical activities that form the framework for materials and energy flows in a product life cycle

Pollution   The unwanted by-product or residual produced by human activity

Pollution Prevention   The reduction of anything that produces an impact on the environment or health risks that is released into the environment

Postconsumer Materials   Any material that has served its intended purpose and has been discarded before any recovery activities

Recycling   The reprocessing, reforming, or in-process reuse of waste materials; this includes collection, separation, and processing

Renewable   Capable of being replaced with a like entity quick enough to meet present, near-term, and future demands

Residual   The remainder; what is left over after a process or use of a product; the wastes remaining after all usable material has been used or recovered

Reuse   The additional use of a component, product, or material after it has been removed from a clearly defined product or process stage even if some cleaning or processing is required

Sustainable   Able to be maintained through time

System Boundaries   Define the extent of system or activities; boundaries define the limits of activities

Usable Life   Measurement of how long a system, process, or product will operate safely in its intended manner

APPENDIX A

**Example
Process Flow Diagram**

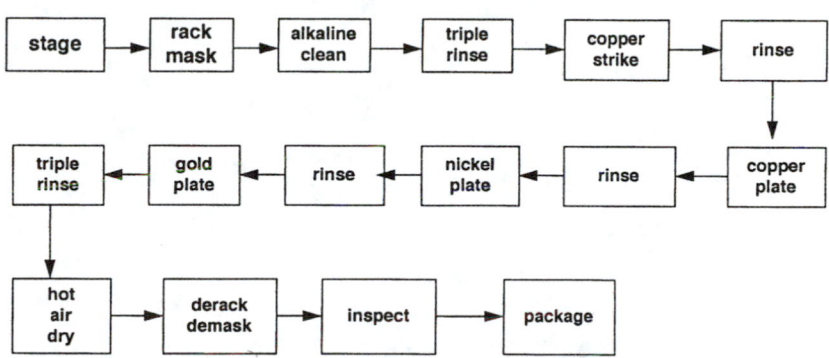

Example of process flow diagram. (From *Waste Minimization as a Strategic Weapon,* Ciambrone, D. F., Ed., CRC/Lewis Publishers, Boca Raton, FL, 1995.)

APPENDIX B

Issues to Consider for Assessing Environmental Impacts

Materials

Amount	Character	Impacts
Type	Virgin	Health and safety
Direct	Recovered (recycled)	Ecological
Product related	Reusable	Residuals
Process related	Resource-based factors	Energy
Indirect	Location	
Fixed capital	Scarcity	
Source	Quality	
Renewable	Regulations	
Lumber	Transportation	
Fishery	Restoration	
Agriculture/	Sustainability	
biomaterials		
Nonrenewable		
Metals		
Nonmetals		
Petrochemicals		

Energy

Amount	Source	Impact/character
Quantity purchased	Renewable	Resource-based factors
Embodied in materials	Wind	Location
Quantity used	Solar	Scarcity
	Hydroelectric	Quality
	Geothermal	Restoration
	Biomass	Impacts
	Nonrenewable	Materials
	Fossil fuel	Residuals
	Nuclear	Ecological
		Health and safety
		Net energy

Residuals		
Type	**Character**	**Environmental fate**
Waterborne	Nonhazardous	Containment
Suspended	Types	Degradability
Biological	Quantities	Bioaccumulation
Emulsified	Hazardous	Mobility
Chemical	Types	Atmospheric
Dissolved	Toxicity	Surface water
Airborne	Concentrations	Biological
Aerosol	Radioactive	Groundwater
Particulate	Type	Treatment impacts
Gas	Amount	Health and safety
Solid	Concentration	effects
Semisolid	Half-life	Materials
Liquid	Potency	Energy
Solid	Medical	Residuals
	Type	Contamination
	Quantity	
	Toxicity	
	Pathogen form	

Ecological factors		
Type	**Stress factors**	**Scale**
Chemical	Sensitive species	Local
Biological	Interrelationships	Regional
Physical	Sustainability	Continental
	Diversity	Global
	Rarity	

Health and safety issues		
Risk to	**Toxicology characteristics**	**Accidents**
Workers	Mortality	Types
Users	Morbidity	Frequency
Community	Exposure routes	Nuisances
	Inhalation	Noise
	Skin contact	Odors
	Ingestion	Sight
	Dose rate	Congestion
	Duration	
	Frequency	
	Concentration	

APPENDIX C

Examples of Energy Storage

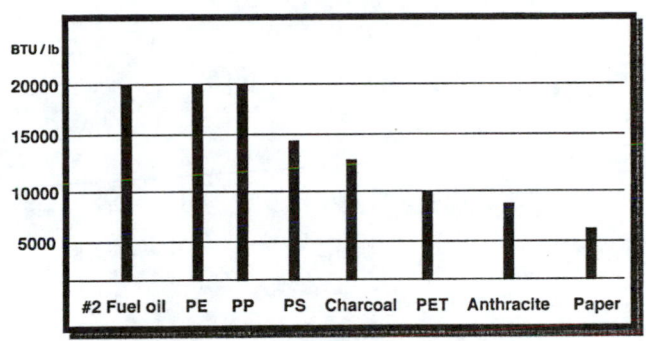

PE = Polyethylene   PET = Polyethylene terephalate

PP = Polypropylene   Anthracite = Hard coal

PS = Polystyrene

APPENDIX D
MAJOR FEDERAL LAWS

CLEAN AIR ACT

Provisions include

Section 3 — National ambient air quality standards
Section 4 — State implementation plan
Section 5 — New source performance standard
Section 6 — Prevention of significant deterioration program
Section 7 — Nonattainment areas
Section 8 — National emissions standards for hazardous air pollutants
Section 9 — Acid rain provisions
Section 10 — New permitting requirements
Section 11 — Mobile source and fuel requirements
Section 12 — Ozone protection

CLEAN WATER ACT

- Grants for construction of treatment works
- National pollution discharge elimination system (NPDES)
- Water quality standards
- New source performance standards
- Toxic and pretreatment effluent standards

COMPREHENSIVE EMERGENCY RESPONSE, COMPENSATION AND LIABILITY ACT (CERCLA) AND SUPERFUND AMENDMENTS AND REAUTHORIZATION ACT (SARA)

- Definitions of hazardous wastes

- National contingency plans
- Liability of responsible parties and financing options for remedial actions

SARA TITLE III

Subtitle A — Emergency response and notification for extremely hazardous substances
Subtitle B — Reporting and notification requirements for toxic and hazardous substances

POLLUTION PREVENTION ACT

- Expand EPA's role in encouraging pollution prevention
- Creates an office in EPA to coordinate all agency pollution prevention activities
- Establishes grant program for training
- Requires the establishment of a clearinghouse to compile and disseminate information on source reduction and to serve as a technology transfer resource
- Mandates that EPA collect data on source reductions, recycling, and treatment of chemicals listed on the TRI reporting forms

RESOURCE CONSERVATION AND RECOVERY ACT (RCRA)

- Responsible to protect human health and the environment, reduce waste and conserve energy and natural resources, and eliminate the generation of hazardous wastes
- Provides for the identification and listing of particularly hazardous wastes for regulation and standards for generators, transporters, and owners/operators of hazardous waste treatment, storage, and disposal facilities (TSDs)
- Establishes recordkeeping, labeling, and manifest systems and proper handling methods
- Requires permits to be granted for treating, storage, or disposal of listed substances; imposes standards applying to financial aspects, groundwater monitoring, minimum technology usage, and closure procedures
- Provides for site inspection and enforcement
- Imposes a requirement on states to compile an inventory describing the hazardous waste storage and disposal sites with each state; encourages each state to take over responsibility for program implementation and enforcement from the federal government

- Establishes guidelines for minimum requirements for solid waste disposal
- Regulates underground storage tanks

TOXIC SUBSTANCE CONTROL ACT (TSCA)

- Provides for the testing of certain substances to determine if they present an unreasonable risk to health or the environment
- Provides for the notification of the EPA before producing a new chemical substance and submits any required test data
- Regulation of substances that present jan unreasonable risk to health and the environment
- Provides for the seizure of substances that are an imminent hazard through civil actions
- Provides for required recordkeeping by manufacturers

FEDERAL INSECTICIDE, FUNGICIDE AND RODENTICIDE ACT (FIFRA)

- Regulates all pesticides, fungicides, and rodenticides and requires that they be registered with the EPA; labeling regulations are included

NATIONAL FOREST MANAGEMENT ACT

- Establishes procedures for sale of forest timber
- Mandates Department of Agriculture to maintain a renewable resource program

OCCUPATIONAL HEALTH AND SAFETY ACT (OSHA)

- Provides for worker health and safety
- Sets standards
- Defines employer duties
- Defines inspection and enforcement
- Hazard communication standard — worker right-to-know laws

SURFACE MINING CONTROL AND RECLAMATION ACT

- Abondoned mine provisions
- Regulates surface coal mining's environmental impact

APPENDIX E
BASIC RECYCLE CODES
FOR PLASTICS

For consumers, the plastics industry indentifies resins by seven basic codes:

 PET (polyethylene teraphthalate) — 1- and 2-liter soda bottles, some foods, sauces, and sundries packaging

 HDPE (high density polyethylene) — gallon milk and juice containers, some food packaging

 PVC (polyvinyl chloride) — food and mineral water bottles

 LDPE (low-density polyethylene) — food product containers and lids, plastic grocery and shopping bags

 PP (polypropylene) — cereal box liners and medical cases

 PS (polystyrene) — deli, cookie and muffin trays, fast food cutlery

 OTHER multiple plastic resins from which layered products such as deli trays and potato and corn chip bags are made

(From *Waste Minimization as a Strategic Weapon*, Ciambrone, D. F., CRC/Lewis Publishers, Boca Raton, FL, 1995.)

INDEX